Designing
Cost-Efficient
Mechanisms

Designing Cost-Efficient Mechanisms

Minimum Constraint Design,
Designing with Commercial Components,
and Topics in Design Engineering

Lawrence J. Kamm

McGraw-Hill Professional & Reference Division offers business audio tapes & CD-ROM products in science and technology. For information & a catalogue, please write to Audio/Optical Media Dept., McGraw-Hill Professional & Reference Dept., 11 West 19th St., NY, NY 10011

McGraw-Hill, Inc.
New York   St. Louis   San Francisco   Auckland   Bogotá
Caracas   Hamburg   Lisbon   London   Madrid
Mexico   Milan   Montreal   New Delhi   Paris
San Juan   São Paulo   Singapore
Sydney   Tokyo   Toronto

STAFFORD LIBRARY
COLUMBIA COLLEGE
1001 ROGERS STREET
COLUMBIA, MO 65216

To my wife, Edith, for more reasons than I care to publish.

Library of Congress Cataloging-in-Publication Data

Kamm, Lawrence J.
   Designing cost-efficient mechanisms: minimum constraint design, designing with commercial components, and topics in design engineering / Lawrence J. Kamm.
     p.   cm.
  ISBN 0-07-033569-9
  1. Machinery—Design. I. Title.
TJ230.K25 1990
621.8'15—dc20   89-29373

Copyright © 1990 by McGraw-Hill, Inc. All rights reserved. Printed in the United States of America. Except as permitted under the United States Copyright Act of 1976, no part of this publication may be reproduced or distributed in any form or by any means, or stored in a data base or retrieval system, without the prior written permission of the publisher.

1 2 3 4 5 6 7 8 9 0 DOC/DOC 9 6 5 4 3 2 1 0

ISBN 0-07-033569-9

The sponsoring editor for this book was Harold B. Crawford, the editing supervisors were Beatrice E. Eckes and Stephen M. Smith, the designer was Naomi Auerbach, and the production supervisor was Dianne L. Walber. It was set in Century Schoolbook by McGraw-Hill's Professional and Reference Division composition unit.

Printed and bound by R. R. Donnelley & Sons Company.

Information contained in this work has been obtained by McGraw-Hill, Inc., from sources believed to be reliable. However, neither McGraw-Hill nor its authors guarantees the accuracy or completeness of any information published herein and neither McGraw-Hill nor its authors shall be responsible for any errors, omissions, or damages arising out of use of this information. This work is published with the understanding that McGraw-Hill and its authors are supplying information but are not attempting to render engineering or other professional services. If such services are required, the assistance of an appropriate professional should be sought.

For more information about other McGraw-Hill materials, call 1-800-2-MCGRAW in the United States. In other countries, call your nearest McGraw-Hill office.

Contents

Preface   xix

Part 1 Minimum Constraint Design (MinCD), Semi-MinCD, and Redundant Constraint Design (RedCD)

Chapter 1. General Description   3

Chapter 2. Degrees of Constraint   5

2.1 Disadvantages and Benefits of RedCD   5
    2.1.1 Disadvantages   5
        2.1.1.1 Part-to-Part Variation   5
        2.1.1.2 Assembly Stresses and Strains   6
        2.1.1.3 Deformation in Normal Service   6
        2.1.1.4 Damage Deformation   6
        2.1.1.5 Thermal Deformation   6
        2.1.1.6 Wear Deformation   7
    2.1.2 Benefits   7
        2.1.2.1 Deformation to Assemble   7
        2.1.2.2 Operating Deformation   7
        2.1.2.3 Load-Spreading Deformation   7
2.2 Theory of MinCD   8
    2.2.1 Axes   8
    2.2.2 Freedoms and Constraints   8
    2.2.3 An Example of Pure MinCD   9
    2.2.4 Degree of Purity   10
    2.2.5 Further Theory   11
    2.2.6 Rules and Principles   11
    2.2.7 Rotary Constraints   12
    2.2.8 Matched Sets   12
    2.2.9 Relative Constraint   13
    2.2.10 Needed Theory   14
2.3 Examples of Bad RedCD   14
        2.3.1.1 Three Bearings on One Shaft   14
        2.3.1.2 Dovetail Slides   14
        2.3.1.3 Bolted Feet   15
        2.3.1.4 Lead Screw   16
        2.3.1.5 Chairs and Tables   17

2.4	Examples of Good RedCD		17
	2.4.1.1	Cylinder Head	17
	2.4.1.2	Flanged Joint and Bolt Circle	17
2.5	Examples of MinCD		17
	2.5.1	The Ubiquitous Tripod	17
		2.5.1.1 Surveyor's Instrument Tripod	17
		2.5.1.2 One-Leg Stool	18
		2.5.1.3 Three-Leg Chairs	19
		2.5.1.4 Kettles	20
		2.5.1.5 Ancient Tripods	20
		2.5.1.6 Tension Tripods	20
		2.5.1.7 Tripod Derricks	20
		2.5.1.8 Surface Plates	21
		2.5.1.9 Machine Tools	21
		2.5.1.10 Jigs and Fixtures	21
		2.5.1.11 Tricycles	22
		2.5.1.12 Trailers	23
		2.5.1.13 Bell Striker	23
	2.5.2	Examples of MinCD in Industry	23
		2.5.2.1 Lathe Chucks	23
		2.5.2.2 Robot Grippers	23
		2.5.2.3 Straight-Line Mechanism	24
		2.5.2.4 Larger Straight-Line Mechanism	25
		2.5.2.5 Large Linear-Motion Mechanism	25
		2.5.2.6 Large Storage and Retrieval Robot	26
		2.5.2.7 Safety Caging	28
		2.5.2.8 Assembly	29
		2.5.2.9 Tandem Shafts	30
2.6	MinCD with Flexible Bodies		30
	2.6.1	Examples of Flexible Body	30
		2.6.1.1 Long Machine Beds	30
		2.6.1.2 Robot Spar	31
		2.6.1.3 Flexible Cart	31
	2.6.2	Classes of Flexible Body	32
		2.6.2.1 Thin	32
		2.6.2.2 Long	32
		2.6.2.3 Large	32
2.7	Load Dividers		32

Chapter 3. Kinds of Constraint 35

3.1	Hard Constraints		35
	3.1.1	Examples of Hard Constraints	35
		3.1.1.1 Point and Surface	35
		3.1.1.2 Ball and Surface	35
		3.1.1.3 Roller and Surface	36
		3.1.1.4 Shaft-and-Sleeve Bearing	36
		3.1.1.5 Ball and Socket	36
		3.1.1.6 Bolted Feet	36
	3.1.2	Examples of Wheel Constraints	36
		3.1.2.1 Single Narrow Wheel	36
		3.1.2.2 Pair of Wheels Tight on a Common Axle	36
		3.1.2.3 Pair of Wheels Loose on a Common Axis	37
	3.1.3	Examples of Wheels on Tracks	37
		3.1.3.1 Two Flanges on One Wheel	37

		3.1.3.2 One Flange on Each Wheel of a Pair	37
		3.1.3.3 Traditional Railroad Wheels	37
		3.1.3.4 V Grooves	37
	3.1.4	Examples of Rotary Hard Constraints	37
		3.1.4.1 Jaw Clutch	37
		3.1.4.2 Splined Shaft	37
		3.1.4.3 U Joint, Splined Shaft, U Joint	38
		3.1.4.4 Independent Rotary Constraint	38
3.2	Centering Constraints		38
	3.2.1	Hard Centering	38
	3.2.2	Soft Centering	39
3.3	Human Constraints		40
3.4	Soft Constraints		42
	3.4.1	Uses for Soft Constraints	42
		3.4.1.1 Shock and Vibration Isolation	42
		3.4.1.2 Oscillation Damping	42
		3.4.1.3 Contact Stress Reduction	43
		3.4.1.4 Scratch and Dent Prevention	43
		3.4.1.5 Pressure Distribution	43
		3.4.1.6 Overtravel Cushioning	43
		3.4.1.7 Separation of Sliding Parts	44
	3.4.2	Seating Forces	45
	3.4.3	Materials Used in Soft Constraints	46
	3.4.4	Effects Used in Soft Constraints	46
		3.4.4.1 Elasticity	46
		3.4.4.2 Hysteresis	46
		3.4.4.3 Viscosity	46
		3.4.4.4 Buoyancy	47
		3.4.4.5 Eddy Currents	48
		3.4.4.6 Magnetic Attraction and Repulsion	48
		3.4.4.7 Gravity	48
		3.4.4.8 Fluid Pressure	49
	3.4.5	Devices Used for Soft Constraint	49
		3.4.5.1 Dashpots	49
		3.4.5.2 Cylinders	50
		3.4.5.3 Motors	51
		3.4.5.4 Voice Coils	51
		3.4.5.5 Clutches and Brakes	51
		3.4.5.6 Shock Mounts	51
	3.4.6	Inflated Constraint Devices	52
		3.4.6.1 Inflated Seal	52
		3.4.6.2 Air Springs	52
	3.4.7	Part Grippers	53
		3.4.7.1 Inflated Tube	53
		3.4.7.2 Expanding O Rings	53
		3.4.7.3 Vacuum Grippers	53
		3.4.7.4 Ventricles	54
		3.4.7.5 Squeeze Valve	54
		3.4.7.6 Air Tube Clutch and Brake	54
		3.4.7.7 Inflatable Actuator	55
3.5	Flexible Constraints		55
		3.5.1.1 Flexures	55
		3.5.1.2 Flexible Couplings	57
		3.5.1.3 Torsion Flexures	58
		3.5.1.4 Suspension Flexures	58

				Page
		3.5.1.5	Bimetal	59
		3.5.1.6	Tape	59
		3.5.1.7	Electrical Flexures	59
		3.5.1.8	Tension Flexures	61
		3.5.1.9	Hoses	61
		3.5.1.10	Flexible Hose and Cable Supports	61
		3.5.1.11	Energy Storage Flexures	62
		3.5.1.12	Balancing Springs	62
		3.5.1.13	Flexible Containers	62
		3.5.1.14	Musical Instruments	62
		3.5.1.15	Clock Crystals	62
		3.5.1.16	Latches	62
3.6	Adjustable Constraints			63
	3.6.1	Reasons for Adjustability		63
	3.6.2	Adjustable Parameters		63
		3.6.2.1	Straightness	63
		3.6.2.2	Squareness, Plumbness, and Levelness	63
		3.6.2.3	Angle	63
		3.6.2.4	Phase	63
		3.6.2.5	Distance	63
		3.6.2.6	Alignment and Parallelism	63
		3.6.2.7	Force and Pressure	63
		3.6.2.8	Electrical Parameters	64
	3.6.3	Adjustment Techniques		65
		3.6.3.1	Adjusting Screws	65
		3.6.3.2	Eccentrics	66
		3.6.3.3	Shims	66
		3.6.3.4	Wedges	66
		3.6.3.5	Phasing Hubs	66
		3.6.3.6	Adhesives and Grout	66
		3.6.3.7	Hammers and Files	66
		3.6.3.8	Part Replacement	66
	3.6.4	Adjustment Measurements		66
	3.6.5	Geometrical Instruments		67
		3.6.5.1	Level	67
		3.6.5.2	Square	67
		3.6.5.3	Angle Scales	67
		3.6.5.4	Plumb Line	67
		3.6.5.5	Linear-Distance Instruments	68
		3.6.5.6	Optical Measuring Systems	68
	3.6.6	Other Instuments		68
		3.6.6.1	Human Touch, Hearing, and Vision	68
		3.6.6.2	Transducers	69
3.7	Variable Constraints			69
		3.7.1.1	Cams	69
		3.7.1.2	Linkages	70
		3.7.1.3	Lead Screws	70
		3.7.1.4	Air Cylinders	70
		3.7.1.5	Hydraulic Cylinders	70
		3.7.1.6	Electric Motors	71
		3.7.1.7	Gears	72
		3.7.1.8	Rack and Pinion	72
		3.7.1.9	Chain, Tape, and Rope	73
		3.7.1.10	Belt and Pulley	73
		3.7.1.11	Servos	74

- 3.8 Friction Constraints — 74
  - 3.8.1 Friction Devices — 74
    - 3.8.1.1 Collets — 74
    - 3.8.1.2 Chucks and Vises — 75
    - 3.8.1.3 Wedges — 75
    - 3.8.1.4 Taper Pins — 75
    - 3.8.1.5 Setscrews — 75
    - 3.8.1.6 Belt and Pulley — 75
    - 3.8.1.7 Friction Variable-Speed Drives — 75
  - 3.8.2 Screw Thread Retention — 75
    - 3.8.2.1 Plastic Inserts — 75
    - 3.8.2.2 Self-Tapping Screws — 75
    - 3.8.2.3 Deformed Threads — 75
    - 3.8.2.4 Tapered Threads — 75
    - 3.8.2.5 Adhesives — 76
    - 3.8.2.6 Lockwashers — 76
    - 3.8.2.7 Locknuts — 76
- 3.9 Self-Aligning Elements — 76
  - 3.9.1.1 Caster — 76
  - 3.9.1.2 Single Gimbal — 77
  - 3.9.1.3 Two Gimbals — 77
  - 3.9.1.4 Two Gimbals (Alternate) — 77
  - 3.9.1.5 Three Gimbals — 77
  - 3.9.1.6 Sliding Gimbals — 78
  - 3.9.1.7 Universal Joints — 79
  - 3.9.1.8 Self-Aligning Linear Ball Bearing — 79
  - 3.9.1.9 Spherical Joint ("Ball Joint") — 79
  - 3.9.1.10 Spherical Bearing Mount — 81
  - 3.9.1.11 Leveling Pads — 82
  - 3.9.1.12 Spherical Bearing — 83
  - 3.9.1.13 Spherical Washers — 85
  - 3.9.1.14 Self-Aligning Roller Bearing — 86
  - 3.9.1.15 Linear Chained Roller Bearing — 86
  - 3.9.1.16 Ball Caster — 87
  - 3.9.1.17 Spline — 87
  - 3.9.1.18 Active Self-Alignment — 91

Chapter 4. Beneficial Non-MinCD — 93

- 4.1 Semi-MinCD — 93
- 4.2 Matched Sets — 93
- 4.3 Finite Area Contacts — 94
- 4.4 MinCD to Semi-MinCD Conversion — 94
  - 4.4.1.1 Conversion of Fig. 2.1 — 94
  - 4.4.1.2 Heavily Loaded Slide — 94
  - 4.4.1.3 Lathe Carriage — 94
  - 4.4.1.4 Bolted Foot — 95
  - 4.4.1.5 Zero-Looseness Hinge — 95
- 4.5 Useful RedCD — 95
  - 4.5.1.1 Large Distributed Load — 95
  - 4.5.1.2 Necessary Deformation — 96
  - 4.5.1.3 Varying Load Distribution — 96
- 4.6 RedCD Components — 97
  - 4.6.1.1 V-Band Fastener — 97

x Contents

	4.6.1.2 Retaining Rings	98
	4.6.1.3 Screw Threads	100
	4.6.1.4 "Piano Hinge"	101
	4.6.1.5 Flanged Joint	101
4.7	Self-Improving RedCD	101
	4.7.1.1 Wearing In	101
	4.7.1.2 Flat Lapping	101
	4.7.1.3 Parabolic Lapping	101
	4.7.1.4 Bearing-Ball Lapping	101
	4.7.1.5 Circle Divider	102
	4.7.1.6 Lead-Screw Lapping	103
	4.7.1.7 Hand Scraping	103
	4.7.1.8 Conical Bearings	104
	Exercises in MinCD, Semi-MinCD, and RedCD	104

Part 2 Designing with Commercial Components

Chapter 5. General Discussion — 107

5.1	Commercial versus Special	107
	5.1.1 Advantages of Commercial Components	107
	5.1.1.1 Development Costs	107
	5.1.1.2 Manufacturing Costs	107
	5.1.1.3 Experience	107
	5.1.1.4 Approvals	107
	5.1.2 Advantages of Your Own Design	107
	5.1.2.1 Suitability	107
	5.1.2.2 Costs	108
	5.1.2.3 Design Integration	108
	5.1.2.4 Independence	108
	5.1.2.5 Management Considerations	108
	5.1.2.6 Combining Ideas	108
5.2	Approved Products	108
	5.2.1 Your Company	108
	5.2.2 Other Organizations	109
	5.2.3 Your Customer	109
5.3	Sources of Information	110
	5.3.1 Your Program of Study	110
	5.3.1.1 Catalogs	110
	5.3.1.2 Advertisements	110
	5.3.1.3 Trade Shows	110
	5.3.2 Purchasing Directories	111
	5.3.3 Manufacturers' Representatives and Salespeople	111
5.4	Big Companies versus Small Companies	111
5.5	Components in This Book	112
5.6	Organization of This Book Section	112
5.7	Breadth and Depth	112
5.8	How to Use This Book	113

Chapter 6. Rotary Motion — 115

6.1	Bearings	115
	6.1.1 Rolling Bearings	115

		6.1.1.1 Ball Bearings	115
		6.1.1.2 Roller Bearings	116
	6.1.2	Bearing Housings	116
	6.1.3	Sliding Bearings	117
		6.1.3.1 Hydrodynamic Lubrication	117
		6.1.3.2 Hydrostatic Lubrication	117
		6.1.3.3 Dry	117
	6.1.4	Flexure Bearings	118
6.2	Spindle Assemblies		118
6.3	Coupling Hubs to Shafts		118
	6.3.1	Interference Couplings	118
	6.3.2	Tapers and Collets	119
6.4	Collars and Retaining Rings		119
6.5	Shafting		119
6.6	Clutches and Brakes		120
	6.6.1	Torque-Generating Effects	121
		6.6.1.1 Dry Friction	121
		6.6.1.2 Lubricated Friction	121
		6.6.1.3 Hydrodynamic Forces	121
		6.6.1.4 Viscous Drag	121
		6.6.1.5 Magnetic Particle	121
		6.6.1.6 Eddy-Current Drag	122
		6.6.1.7 Hysteresis Drag	122
		6.6.1.8 Positive Engagement	122
		6.6.1.9 Generators and Motors	122
	6.6.2	Control Effects	122
		6.6.2.1 Electricity	123
		6.6.2.2 Compressed Air	123
		6.6.2.3 Hydraulics	123
		6.6.2.4 Centrifugal Force	123
		6.6.2.5 Torque	123
		6.6.2.6 Angular Position	124
		6.6.2.7 Human	124
6.7	Rotation Transmission		124
	6.7.1	Shaft Couplings	124
	6.7.2	Gears	125
	6.7.3	Gearless Speed Reducers	125
	6.7.4	Friction Drives	125
		6.7.4.1 V Belts	125
		6.7.4.2 Multi-V Belts	125
		6.7.4.3 Flat Belts	126
		6.7.4.4 Tooth Belts	126
		6.7.4.5 Round Belts	126
	6.7.5	Chains	126
	6.7.6	Indexing Drives	127
	6.7.7	Variable-Speed Drives	127
		6.7.7.1 Variable-Speed Motors	127
		6.7.7.2 Motor and Slip Clutch	127
		6.7.7.3 Friction	128
		6.7.7.4 V Belt	128
		6.7.7.5 Hydraulic	128
		6.7.7.6 Gearshift	128

xii   Contents

Chapter 7. Linear Motion — 129

- 7.1 Bearings, Wheels, and Tracks — 129
  - 7.1.1 Roller and Track Matched Sets — 129
    - 7.1.1.1 Round Tracks — 129
    - 7.1.1.2 Nonround Tracks — 130
    - 7.1.1.3 Roller-Bearing Systems — 130
  - 7.1.2 Complete Matched Sets — 130
- 7.2 Wheels — 131
  - 7.2.1 Wheels for Flat Paths — 131
- 7.3 Wheel Steering — 132
- 7.4 Wheel and Track Matched Sets — 132
- 7.5 Hydrostatic Sliding Bearings — 132
- 7.6 Lead Screws and Nuts — 133
- 7.7 Belts, Chains, and Ropes — 133

Chapter 8. Power — 135

- 8.1 Available Forms of Power — 135
- 8.2 Power Sources — 135
  - 8.2.1 Electricity — 135
    - 8.2.1.1 Motors — 136
    - 8.2.1.2 Other Electric Actuators — 137
    - 8.2.1.3 Heaters — 137
    - 8.2.1.4 Electrical Controls — 137
    - 8.2.1.5 Wiring Devices — 138
  - 8.2.2 Hydraulics — 138
    - 8.2.2.1 Cylinders — 138
    - 8.2.2.2 Motors — 139
    - 8.2.2.3 Control Devices — 139
    - 8.2.2.4 Plumbing — 139
  - 8.2.3 Pneumatics — 140
  - 8.2.4 Explosives — 141
  - 8.2.5 Springs — 141
  - 8.2.6 Flywheels — 142
  - 8.2.7 Heat Engines — 142
  - 8.2.8 Fuel Burning — 142
  - 8.2.9 Human Power — 142

Chapter 9. Other Components — 145

- 9.1 Semifinished Materials — 145
- 9.2 Structural Systems — 146
- 9.3 Enclosures — 146
- 9.4 Machine Modules — 147
- 9.5 Fasteners — 148
  - 9.5.1 Threaded Fasteners — 148
    - 9.5.1.1 Data and Specifications — 148
    - 9.5.1.2 Threaded Inserts — 148
    - 9.5.1.3 Sheet-Metal Nuts — 148
    - 9.5.1.4 Other Forms — 148
  - 9.5.2 Thread Locking — 149
    - 9.5.2.1 Lockwashers — 149

Contents   xiii

9.5.2.2	Locknuts	149
9.5.2.3	Lock Wire	150
9.5.2.4	Castellated Nuts	150
9.5.2.5	Insert Nuts	150
9.5.2.6	Insert Bolts	150
9.5.2.7	Deformed Nuts	150
9.5.2.8	Adhesives	150
9.5.3	Nonthreaded Fasteners	150
9.5.3.1	Rivets	150
9.5.3.2	Other Fasteners	151
9.5.3.3	Latches	151
9.6	Vibration and Shock Absorbers	151
9.6.1	Shock Mounts	151
9.6.2	Shock Absorbers	152
9.7	Springs	152
9.8	Lubrication	153
9.9	Seals and Guards	154
9.10	Sensors and Displays	154
9.10.1	Parameters	155
9.11	Sequence Controls	156
9.11.1	Timers	156
9.11.2	Drum Controllers	156
9.11.3	Relay Circuits	156
9.11.4	Programmable Controllers (PLC)	157
9.11.5	Computers	157
9.11.6	Nonelectrical Controllers	157
9.12	Tooling Components	158
9.13	Permanent Magnets	158
9.14	Lamps	158
9.15	Nameplates	158
9.16	Pumps and Blowers	159
9.17	Miscellaneous	159
	Exercises in Design with Commercial Components	159

Part 3   Topics in Design Engineering

Chapter 10. Designing with Uncommon Manufacturing Processes   163

Chapter 11. Manufacturing Engineering   167

11.1	What Is Manufacturing Engineering?	167
11.1.1.1	Standard Machines	167
11.1.1.2	Special Machines for Sale	168
11.1.1.3	Special Machines for Your Company	168
11.1.1.4	R&D	168
11.1.1.5	Tool Design	168
11.1.1.6	Planning and Scheduling	168
11.1.1.7	Maintenance	168
11.2	Suggestions	169
11.2.1.1	Risk Responsibility	169
11.2.1.2	Technician Work; Engineering Work	169
11.2.1.3	Motivation	169
11.2.1.4	Offices	170
11.2.1.5	Education	170

Chapter 12. Optimum Level of Mechanization and Automation — 171

- 12.1 Classification — 171
  - 12.1.1 Fully Automatic — 172
  - 12.1.2 Powered Machines with Human Control — 172
  - 12.1.3 Combination Human and Automatic — 173
  - 12.1.4 Human Work with Power Tools — 173
  - 12.1.5 Human Workers with Special Hand Tools — 173
- 12.2 Assembly Kits — 174
- 12.3 The Benefits of Automation — 174
- 12.4 Justifying the Cost of Automation — 174
- 12.5 Policy Questions — 175

Chapter 13. Robots — 177

- 13.1 History and Myth — 177
- 13.2 Robot Reality — 178
  - 13.2.1 End Effectors — 178
    - 13.2.1.1 Fabricating Tools — 178
    - 13.2.1.2 Material-Handling Tools — 178
    - 13.2.1.3 Sensors — 178
    - 13.2.1.4 Military Components — 179
    - 13.2.1.5 Quick-Change Grippers — 179
- 13.3 Robot Control — 179
  - 13.3.1 Point-to-Point Robots — 179
  - 13.3.2 Continuous-Path Robots — 180
  - 13.3.3 Human Remote Control — 180
- 13.4 Robot Mechanisms — 180
  - 13.4.1 Linear versus Rotary Axes — 181
    - 13.4.1.1 Errors — 181
    - 13.4.1.2 Flexibility — 181
    - 13.4.1.3 Inertia — 181
    - 13.4.1.4 Geometry Computation — 181
- 13.5 Cartesian Robots — 182
  - 13.5.1.1 Accuracy — 182
  - 13.5.1.2 Flexibility — 182
  - 13.5.1.3 Inertia — 182
  - 13.5.1.4 Position Computation — 182
  - 13.5.1.5 Control — 182
  - 13.5.1.6 Modularity — 183
  - 13.5.1.7 Branching — 183
  - 13.5.1.8 System Configuration — 184
- 13.6 Safety — 184
- 13.7 Cartesian Robot Configurations — 184
  - 13.7.1.1 Bridge Crane — 184
  - 13.7.1.2 Half Bridge — 184
  - 13.7.1.3 Vertical Bridge — 184
- 13.8 Programming — 185
- 13.9 Accuracy versus Repeatability — 186
- 13.10 Conversion from Task to Task — 187
- 13.11 Variation within Task — 187
- 13.12 Money — 187

13.13	Humans versus Robots		188
	13.13.1.1	Task-to-Task Conversion	188
	13.13.1.2	Capital Cost	188
	13.13.1.3	Multiple Tasks	188
	13.13.1.4	Mobility	188
	13.13.1.5	Expanded Scope	188
	13.13.1.6	Bad-Part Rejection	188
	13.13.1.7	Task Modification	188
	13.13.1.8	Dexterity	188
	13.13.1.9	Speed	188
	13.13.1.10	Maintenance	189
	13.13.1.11	Technology	189
13.14	Disadvantages of Human Workers		189
	13.14.1.1	Work Uniformity	189
	13.14.1.2	Unions	189
	13.14.1.3	Fatigue	189
	13.14.1.4	Conflicts	189
	13.14.1.5	Absenteeism	189
	13.14.1.6	Injuries	189
13.15	Economic Justification		189
13.16	Task Size and Force		190
13.17	Abuse Resistance		190
13.18	The Future		190

Chapter 14. Robot Grippers — 191

14.1	Methods of Gripping		191
	14.1.1	Friction	191
	14.1.2	Vacuum	192
	14.1.3	Electromagnets	195
	14.1.4	Special Gripping Devices	196
14.2	Environmental Limitations		198
14.3	Gripper Actuation		199
14.4	Misalignment		200
	14.4.1	Passive Self-Alignment	201
	14.4.2	Active Self-Alignment	203

Chapter 15. Selecting Power Forms — 205

15.1	Forms of Power		205
	15.1.1	Electricity	206
	15.1.2	Pneumatics	207
	15.1.3	Hydraulics	208
		15.1.3.1 Force	208
		15.1.3.2 Power	208
		15.1.3.3 Incompressibility	208
	15.1.4	Vacuum	209
	15.1.4	Vacuum	209
	15.1.5	Combustion Engines	209
	15.1.6	Explosives	210
	15.1.7	Human Muscle	210
	15.1.8	Heat	211

xvi Contents

    15.1.9 Sunlight   211
    15.1.10 Winds   211
    15.1.10 Wind   211
    15.1.11 Gravity   212
    15.1.12 Elasticity   212
    15.1.13 Inertia   213
    15.1.14 Utility Water   213
    15.1.15 Nuclear   214

Chapter 16. Backlash   215

Chapter 17. Hype   223

Chapter 18. Product Deterioration   225

  18.1 Spontaneous Deterioration   225
  18.2 Attacks during Shipment   225
  18.3 Environmental Attacks   226
  18.4 Wear   227
  18.5 Abuse   227
  18.6 Design and Manufacturing Errors   227
  18.7 Modification by User   227
  18.8 What Can You Do about It?   227

Chapter 19. Electrical and Mechanical Technologies: Competition and Cooperation   229

  19.1 History   229
  19.2 Electrical Takeovers   230
        19.2.1.1 Motors   230
        19.2.1.2 Variable-Speed Motors   230
        19.2.1.3 Instrumentation and Control   230
        19.2.1.4 Programming   230
        19.2.1.5 Computing   230
  19.3 Mechanical Instruments   231
  19.4 Uses for Electricity in Machines   232
        19.4.1.1 Transmission of Mechanical Power   232
        19.4.1.2 Heating   232
        19.4.1.3 Information   232
        19.4.1.4 Miscellaneous   232
  19.5 The Future of Mechanism   232
  19.6 Fields of Mechanism   233
      19.6.1 Engines and Turbines   233
      19.6.2 Moving Matter   234
          19.6.2.1 Vehicles   234
          19.6.2.2 Material Handling   234
          19.6.2.3 Manufacturing   234
          19.6.2.4 Domestic Appliances   235
          19.6.2.5 Military Devices   235
          19.6.2.6 Surgical Devices   235
      19.6.3 Structures   235
  19.7 Conclusion   235

Chapter 20. References and Bibliography   237

20.1 Books on Design   237
20.2 Books on Information Sources   238
20.3 Journal Articles on Design   238
20.4 Handbooks   238
    20.4.1 Handbooks on Mechanism Design   238
20.5 Technology Encyclopedias   239
20.6 Specifications and Standards   239
    20.6.1 Commercial Sources for Military Specifications   240
20.7 Journals with New-Product Announcements   240
20.8 Directories   240
    20.8.1 Purchasing Directories   240
    20.8.2 Business Information about Vendors   241
20.9 Mail-Order "Department Store" Catalogs   241

Index   243

Preface

This is a book of ideas and not of calculations. There are many other books and courses which teach you analytic design and the computer programming to help you do it. This book describes some uncommon approaches to *qualitative* design up to the point where calculations are needed for *quantitative* design, i.e., for sizing the parts.

This book is intended for senior engineering students, for practicing mechanical engineers, and for nondegreed mechanical designers. *It contains material not usually taught and little known.* Its treatment is nonmathematical, so it should be useful to mechanism designers of all lengths of training and experience.

In addition to the material indicated by the chapter headings, I have included a number of design ideas which I have found useful and which I hope may be useful to you, and a number of anecdotes which illustrate some of the book material as it appeared in the real world. Part 3 of this book is devoted entirely to design and manufacturing ideas.

Part 1 concerns minimum constraint design (MinCD), semi-MinCD, and redundant constraint design (RedCD).

MinCD provides zero binding and zero looseness of moving parts. It provides zero stresses from assembling and installing stationary assemblies. MinCD provides all these benefits despite loose manufacturing tolerances and semiskilled assembly labor. MinCD provides major cost reductions in both product design and manufacturing and major increases in both reliability and maintainability.

The principles of MinCD and semi-MinCD apply to all mechanisms from sensitive instruments such as gravity meters to massive machinery such as earth movers.

MinCD is known to few mechanical designers as a systematic process although some good designers use it intuitively. I was first introduced to MinCD many years ago by Ref. 1 of Chap. 20 of this book, and I have been practicing, developing, and teaching it ever since, with great advantage to my companies, to my clients, and to myself.

Understanding minimum constraint design will increase your insight (your "gut feel") into some aspects of the behavior of mechanisms and structures and will increase your skill as a designer by giving you both theory and examples.

Part 2 of this book will help you to design mechanisms of maximum value and minimum cost by using commercial components instead of designing special devices. Much of electrical, pneumatic, and hydraulic design consists of connecting commercial components together to form functioning circuits. Analogously, much of mechanism design can be the joining of commercial components into functioning mechanisms. Even if you decide to design a special, existing technology will give you a fund of ideas; you need not reinvent the wheel.

Commercial components are of particular value for small-quantity production where the cost of engineering and manufacturing a special may be prohibitive; many machines are produced in quantities of one. Yet in even the smallest quantities commercial components cost only the price of the individual components you buy, which are mass-produced as products by others.

This book contains names of companies whose catalogs have a high educational content; these catalogs are textbooks on their components and their components' uses. I recommend these catalogs for your technical library as extensions of this book. Do *not* assume that the sources of these teaching catalogs are either the only or the best vendors of the components described. There is a huge marketplace of competing products and vendors and there is no Consumers Union to compare benefits and make recommendations. There *are* qualified-products lists published by your customer, your own company, and certain testing laboratories; the Bibliography will help you find some of them. *One of your continuing career responsibilities is to study components and vendors and form your own opinions of their values.*

There are no names of component vendors other than the manufacturers who publish these instructive catalogs or whose unique or typical products are described in the text. Lists of vendors are published in industrial directories; in the References is a list of such directories. These directories are very large books, and some are essential components of your personal or company library. They not only guide you to manufacturers but give business information about those manufacturers. Browsing through them may also guide you to components you did not know existed.

The References give the names of some multiple-product "department store" catalogs which can be mother lodes of unusual components and materials.

Either in choosing or in designing components you must comply with a multitude of specifications. The References include sources of many of the specifications with which you must live.

"Component" and "system" are relative terms; every company combines components into systems. To a mechanism designer a bearing is a component and a machine is a system. To an admiral an aircraft carrier is a component and a fleet is a system. All manufacturers *buy* components and *sell* systems. "System" is a more prestigious word, so it is sometimes overused. In this book, components are assembled into machines.

Part 3 is a set of essays on mechanical design. Some make the difference between a theoretically operable machine and a machine which is profitable, long-lived, and "friendly" to its manufacturer, user, and maintainer. Others provide ideas and information not usually included in college courses. All are intended to enhance your knowledge and therefore your career.

The Bibliography lists references for your library and recommendations for your regular reading.

The word "machine" has become so generalized that even an electronic computer is referred to as a machine. I use the word "mechanism" to indicate the mechanical portion of a machine which may have electrical and other nonmechanical portions.

I believe that this book is an original approach to its subjects, or it would not have been written. However, a penalty for being early is errors, omissions, and controversy. Please send me your suggestions and criticisms; I will gratefully consider them for the next edition, and I will list your name with the names of others whose suggestions and criticisms are so used. Write to me, L. J. Kamm, care of Science and Technology Group, McGraw-Hill, Inc., 11 West 19th Street, New York, New York 10011.

Lawrence J. Kamm
San Diego

ABOUT THE AUTHOR

Larry Kamm's formal education includes a B.S.E.E. from Columbia University in 1941 and an M.E.E. from Brooklyn Polytechnic Institute in 1946. He has Professional Engineer licenses from New York, Maryland, and California, and is a Registered Patent Agent and a Certified Manufacturing Engineer (SME).

He is a member of Sigma Xi and the Institute of Electrical and Electronic Engineers (IEEE), and, when he worked in those fields, of the American Rocket Society (now the American Institute of Aeronautics and Astronautics, AIAA), the Society of Manufacturing Engineers (SME), and the Robot Institute.

Kamm has invented and designed mechanical and electro-mechanical devices of great diversity. These include robots, numerical controls, computer peripherals, space vehicles and components, simulators, a mail sorter memory, a heart-lung machine, automatic test-and-sort and other manufacturing equipment, transducers, switchgear, and engine components. He holds 38 patents, issued and pending.

He has published or presented 26 papers and much trade literature. He is the author of McGraw-Hill's *Successful Engineering*.

Kamm has worked as an employed engineer in both small and large companies, as an entrepreneur (Numerical Control Corp., Devonics, Inc., Typagraph Corp., Mobot Corp.), as a teacher and author in design theory and practice, and as a consulting engineer and inventor, which is his present activity. He lives in San Diego.

Part 1

Minimum Constraint Design (MinCD),

Semi-MinCD, and

Redundant Constraint Design (RedCD)

Chapter 1

General Description

When you do minimum constraint design (MinCD), you support and guide each body only at points, and at as few points as possible to get the desired performance. If you do so, you will achieve zero looseness and zero binding of moving parts; you will achieve assembly of fixed parts without strains or rework; and you will do so despite loose manufacturing tolerances and semiskilled assembly labor. You will minimize the manufacturing cost of your mechanism, you will make it more reliable, you will make it easier to disassemble and reassemble, and you will make it easier to maintain.

These are big promises; let's see if I can deliver.

Compare a three-legged stool with a four-leg chair. The stool cannot wobble no matter how uneven the leg lengths or how rough the floor, but the chair will always wobble because of unevenness of leg lengths and imperfect floor flatness. (When you sit on an accurately made chair on a normal floor, it does not wobble because your weight deforms the chair until the shortest leg touches the floor.) *The stool has minimum constraint design (MinCD), and the chair has redundant constraint design (RedCD).*

This book is not intended to advocate MinCD and condemn RedCD. It will try to give as fair and accurate a description of the benefits and disadvantages of each as it can. For example, in the comparison of stool versus chair, the chair is more resistant to overturning than the stool; in fact, current practice for office tilting chairs is to provide *five* feet (usually with casters) to get almost as much tip resistance as a full circular base.

There are certain *disadvantages* to pure MinCD such as high stress concentrations at its point supports. This disadvantage may be overcome by changing a constraint *point* to a small constraint *area* within which the load is spread. This practice is called semi-MinCD and will be discussed at some length in Chap. 4.

There are many kinds of constraint. Some are hard, some are soft in different ways, some are fixed in space and time, and some vary in space and time. Kinds of constraint will be described in Chap. 3.

Most of those to whom I have taught MinCD became enthusiastic about it. I hope you will enjoy this presentation and will profit from it as much as they have.

Chapter 2

Degrees of Constraint

2.1 Disadvantages and Benefits of RedCD

2.1.1 Disadvantages

There would be no reason to think about MinCD if there were not some disadvantages of conventional RedCD.

If you have, with your own hands, ever helped to assemble hardware defined by your own drawings, you found parts that did not go together without difficulty, unexpected amounts of friction, moving parts which were too loose at some portions of their travel and too tight at others, and tightness or looseness changes after disassembly and reassembly.

In anticipation of these problems you do dimension studies and provide tolerance limits, clearances, and adjustments to make the finished mechanism go together easily and work well. Sometimes you call for selective assembly. Sometimes you call for matched manufacturing of mating parts (e.g., match drilling). Sometimes you provide adjustments and sometimes for adjusting and then doweling an assembly to lock it into the adjusted position.

What are some of the sources of mismatched assembly and uneven motion?

2.1.1.1 Part-to-part variation. The combination of dimensions and tolerances on manufacturing drawings is supposed to assure that any set of parts made in accordance with the drawings will fit together and work together properly. Yet if you were to put a computed tolerance on everything which might possibly affect the performance of your design, such as flatness of sheets and plates and straightness and roundness of holes, your engineering, manufacturing, and inspection costs would be intolerable. So you take some reasonable risks with "standard tolerances" and "good manufacturing practice."

You win some; you lose some.

Oversize fastener holes make it easier to assemble parts, but the resulting dimensional variations of the assembly include the amount of oversize. Assembly fixtures help to reduce assembly misalignment in manufacturing, but they mask future trouble if the parts must be disassembled and reassembled as part of a shipping sequence or during maintenance. Doweling after adjustment assures that reassembly is in the same relative position but makes the disassembly and reassembly more difficult, it costs to do it, the making of the dowel holes cuts through surface finish, and you lose interchangeability of parts.

2.1.1.2 Assembly stresses and strains. Regardless of your dimensions and tolerances and of good manufacturing practice, it is sometimes necessary to force parts in order to line up matching features for assembly. In the auto industry a mallet is not an unknown assembly tool. Forcing may cause both elastic and inelastic residual stresses.

In the real world you will sometimes be challenged, "These parts are out of tolerance, but it will be expensive and time-consuming to replace them; can we use them anyway?" It is hard to "just say no" in the face of a frustrated manufacturing manager.

2.1.1.3 Deformation in normal service. Foundations may settle. Parts may creep inelastically under heavy loads, transient peak loads, and high temperature. Parts deform elastically under the weight of the machine and under working loads and overloads. Parts may shift within the clearance of fastener holes. Such deformations cause stresses in RedCD mechanisms and structures.

2.1.1.4 Damage deformation. Mechanisms in service are subject to inelastic deformation from overload and impact, i.e., from damage. Such deformation may cause looseness or binding of moving parts. The ability to survive minor damage and still work is a major advantage of MinCD mechanisms.

2.1.1.5 Thermal deformation. Temperature changes will expand and contract your machine. If it has parts of different materials, there will be differential expansion and contraction, which may cause stresses. If your machine is large and requires high accuracy to perform properly, such as a large jig borer, temperature gradients may cause unacceptable distortions. The temperature of your machine is not always at steady state. A sudden change of the environmental temperature will change the temperature of the outer parts before the temperature of the inner parts can follow.

2.1.1.6 Wear deformation.
Sliding surfaces, including journal bearings, will wear in when new and will continue to wear if not adequately lubricated. Even ball bearings have some internal sliding of the balls on their races. Wheels wear both from normal rolling and from skidding. Tools, including part grippers, wear from sliding contact with workpieces. *Wear can cause part displacement with consequent looseness or binding in RedCD mechanisms.*

Wear is not all bad; sliding parts wear off their high spots ("asperities") first and get to slide more smoothly. For example, engines "wear in" ("break in" overstates the case). Section 4.7 describes manufacturing processes which use self-improving wear.

2.1.2 Benefits

2.1.2.1 Deformation to assemble.
Part deformation is not always bad. A flanged joint in piping or in large parts of a jet engine joined by a bolt circle benefits from the deformation of the flanges as they are drawn together by multiple (i.e., redundant) fasteners.

Large thin wall cylinders get out of round, but when they are joined, they deform toward each other as the connection is drawn up.

2.1.2.2 Operating deformation.
Flexure and torsion pivots (Sec. 3.5) and springs function by elastic deformation.

Instrument diaphragms, Bourdon tubes, spring-loaded instrument movements, many musical instruments, balance-wheel watches, and quartz crystals use their calibrated elastic deformation to do their work. (The quartz crystals and the musical instruments use their ratio of inertia to stiffness to establish a natural frequency.)

Rupture disks and shear pins use their calibrated failure loads to perform.

2.1.2.3 Load-spreading deformation.
Point constraints used in MinCD designs cause elastic or inelastic deformation in both parts at the points. This deformation spreads the point loads over small areas so that the stresses are reduced to the capacity of the parts.

Rolling balls, rollers, and wheels deform elastically at their regions of contact with their tracks to provide finite contact areas with finite stresses. This is true not only for pneumatic automobile tires but for hardened-steel rolling bearings whose balls or rollers and their races have nominal "points" and "lines" of contact. Gear teeth do the same at their lines of contact. (The stresses at such contacts are called Hertzian stresses and are described in Ref. 26.)

Sections 4.1 through 4.4 are discussions of compromises with pure MinCD and Secs. 4.5 through 4.7 are further discussions of useful RedCD.

2.2 Theory of MinCD

2.2.1 Axes

Any position or uniform motion of any rigid body can be resolved into exactly six component axes, three linear and three rotary. In Cartesian coordinates the linear components can be called X, Y, Z, and the rotary (= angular = orientation) components can be called R_x, R_y, R_z (for rotations about the X, Y, Z axes). Greek letters are often used for rotations.

It is sometimes easier to use words related to your machine instead of the academic X, Y, Z. For example, in a machine through which a web of material flows, such as paper through a copier, you might use "web flow," "transverse," "cross web." I find that the airplane and ship words "pitch," "roll," "yaw" applied to an imaginary airplane or ship attached to the part are clearly understood descriptions of the rotary freedoms of the part.

When I was selling my Cartesian robots, I found that it was possible to describe motions over the telephone with no confusion whatever if I referred to "east or west," "north or south," "up or down," "roll left or right," "pitch up or down," "yaw left or right."

2.2.2 Freedoms and constraints

Each component of motion is called a "degree of freedom," or a "freedom." Something that prevents a freedom is called a "constraint" or a "locator." Several of the following chapters deal with constraints.

One constraint (or locator) will prevent motion on one axis (linear or rotary). Two constraints will prevent motion on two axes, and so on. *Six constraints prevent all motion.* Put differently, five constraints *permit* one freedom, four constraints *permit* two freedoms, and so on. The constraint may be unidirectional in itself but is usually loaded with a "seating force" (see below); so it acts as a bidirectional locator.

One point of contact can provide two or more constraints. For example a chair leg provides one linear constraint vertically and, by friction, provides two linear constraints horizontally (X and Y). If it is a pointed leg (i.e., does not have a broad foot with an area of contact), it provides no rotary constraint at all except in combination with the linear constraints of other legs separated from it by moment arms.

Degrees of Constraint 9

Minimum constraint design (MinCD) provides only the minimum number of constraints needed to permit the freedoms desired and no other freedoms.

2.2.3 An example of pure MinCD

Figure 2.1 illustrates two parts joined by MinCD. This is not just a theoretical example; I have used it in a successful device which I will describe.

Block *A* rests on block *B*. Block *B* has conical hole 1, V-shaped

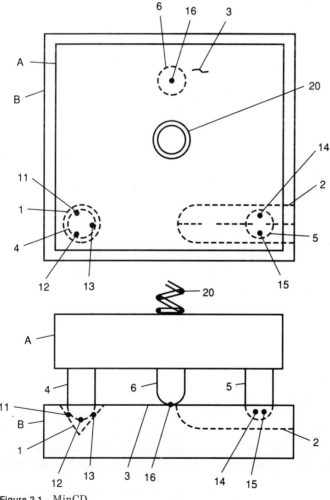

Figure 2.1   MinCD.

groove 2 approximately aligned with conical hole 1, and flat area 3. All tolerances are loose.

Block A has three legs, 4, 5, 6, each ending in approximate hemispheres and resting on block A within loose tolerances.

Let leg 4 rest in conical hole 1. Since neither the leg nor the hole is a mathematically perfect feature, they will touch at three points 11, 12, 13. We now have three constraints, and block A can no longer move in X, Y, or Z, but it can still roll, pitch, and yaw with the end of leg 4 as a center.

Now seat leg 5 in groove 2. It will touch the two sides of the groove at points 14, 15. We now have five constraints and one freedom, which is rotation about the line between the ends of legs 4 and 5.

Now rotate block A about that line until leg 6 touches block B at point 16. Presto: six constraints and zero freedoms.

Note that we are dependent on a force either from gravity or from spring 20 to maintain the constraint points in contact. This force is not a constraint because it is not rigidly oriented or located relative to the blocks. Such a force is called a "seating" or "restoring" force; it is not a constraint since it does not establish position on any of the six axes by itself. Among the sources of seating forces are springs, gravity, centrifugal force, fastener force, air or hydraulic pressure, and magnetic or electro-magnetic force.

I told you that I had used the exact configuration of Fig. 2.1. My problem was to position a removable hot solder pot, with great accuracy, in a location difficult to reach, in an automatic-assembly machine, and to do so frequently and quickly with a semiskilled worker. Block A was the solder pot (with a long insulated handle), and block B was fixed to the machine. Placing the pot in position gave you a strange feeling: one moment it was free in your hand, and the next it was solidly in position with no looseness whatever. Lifting it out was just as easy: zero looseness, zero binding.

This example has hard point contact constraints and a spring seating force. There are many other kinds of constraints, and Chap. 3 will deal with them.

2.2.4 Degree of purity

In the example, leg 4 rests in a conical hole on three constraint points whose positions are determined by part tolerances rather than by design. In a "purer" design the cone-shaped hole is a three-sided pyramid-shaped hole, and there is one constraint point on each pyramid face. In real-world practice the cone is cheaper and just as good. This is a first example of semi-MinCD.

The degree-of-purity question will arise frequently in this book and in your own use of MinCD. Semi-MinCD (Chap. 4) deals with compromises with purity in the interest of lower cost, lower stress, or other considerations. It is important for you to develop your judgment about the value of purity and partial purity in each particular case and about how much you are justified in spending to achieve it.

2.2.5 Further theory

In your study of structures you learned of "statically determinate" and "statically indeterminate" structures. Statically *determinate* structures have support forces and internal loads which can be computed by the equations of statics and are independent of elastic strains in the parts. Statically *indeterminate* structures have loads in some parts which depend in part on the strains of other parts. MinCD constraint forces are statically determinate; RedCD constraint forces are statically indeterminate.

2.2.6 Rules and principles

Although six constraints position a part, not any six will do. For example, consider a sphere resting on six fixed supports. Not only is the sphere still free to rotate in any direction, but the six supports are redundant and, because of tolerances, the sphere will actually touch only three of them unless it (or the supports) is deformed.

Consider a constraint to be a force vector through its point of contact, normal to the surface contacted, and friction-free. (We will talk about real-world friction later.)

If three constraint vectors converge at a point, the part is fully constrained in displacement but is completely unconstrained in orientation. Remember that sphere, above.

For each constraint vector which does not pass through that point, there is one orientation constraint.

A test for overconstraint is to imagine the part lifted off the constraint points one at a time, against the seating force, and sliding along the other constraint points. If this can be done without either lifting off or crushing one or more of the other constraints, the part is not overconstrained. Try this on Fig. 2.1.

"Stability" is a measure of how far the lifting-off process can be carried before the seating-force vector crosses a line between two position constraints and tumbles the part. The stability of a three-legged MinCD stool is less than that of a four-leg RedCD chair and still less than that of a five-caster office chair; the stability of a tricycle car is less than that of a four-wheel car (which shows up quite forcefully when cornering and braking). Greater stability can be a good reason for going to RedCD.

The more widely spaced the constraint vectors, i.e., the longer the

12 Constraint Design

moment arms joining them, the greater is the stability. In many cases, the greater the stability, the less the forces on the constraint points.

2.2.7 Rotary constraints

We have been considering constraints which exert linear forces such as a point touching a surface. There are also purely rotary constraints such as are shown in Fig. 2.2. The sequence of components—U joint, splined joint, U joint—transmits torque constraint between members A and B but does not constrain linear motion between A and B. You will recognize this combination as the coupling between a car's transmission and its rear-axle differential. It constrains the differential input to always have the same angular position as the transmission output despite the motions of the rear axle relative to the chassis. In Chap. 3 I will show you how I used this combination to deliver pure torque in a precision automatic thread-gauging instrument (Fig. 3.47).

2.2.8 Matched sets

A "matched set" is a RedCD assembly of parts whose constraint redundancies can be ignored because of some combination of high manufacturing accuracy, elastic or plastic deformation, wear-in, lubricant fill, or other effects. Examples of matched sets are:

1. A shaft and two accurately aligned sleeve bearings
2. A sphere and its closely fitting spherical seat

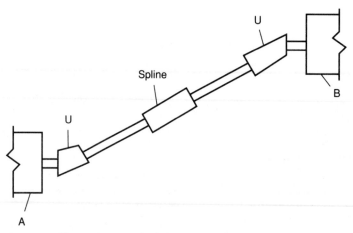

Figure 2.2 Pure rotary constraint.

3. A cylinder in a closely fitting cylindrical seat
4. A cylindrical wheel or roller having very close to line contact with its track
5. A pair of flat areas allowed to self-align
6. A rotation ball bearing
7. A linear ball bearing and its track
8. An accurately made gearbox

You can rely on matched sets to avoid adding parts to achieve MinCD purity. *Semi-MinCD is largely a reliance on matched sets.*

A matched set may be either fixed or adjustable. Figure 2.3 shows an adjustable matched set I have used in a number of mechanisms. It comprises a pair of rollers on a cold-rolled rectangular steel bar used as a rail. Steel bars have one tolerance from bar to bar and a much smaller tolerance from point to point along the same bar. In this adjustable matched set the space between the rollers is adjusted by an eccentric shaft on one of the rollers. (I used commercial crowned cam followers, which can be bought with the eccentric as a standard feature.) Many of the commercial components in Part 2 are matched sets and can be used in otherwise MinCD designs.

2.2.9 Relative constraint

We have been considering constraints between ground (the fixed earth) and a constrained body. The same relationships exist between

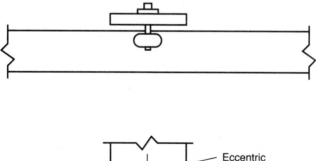

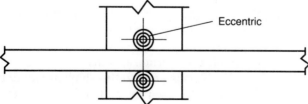

Figure 2.3 Adjustable matched set.

bodies both of which are in motion relative to ground. Examples are the relationship between rotating shafts joined by a flexible coupling and the universal joints and spline in Fig. 2.2.

2.2.10 Needed theory

There is an opportunity here for a more rigorous mathematical analysis of MinCD and a complete set of formal theorems. (A thesis subject, perhaps?)

In any particular case, of course, the forces and torques on the constraints can be computed from the basic equations of statics: total force along each axis equals zero; total moment around each axis equals zero. However, some new design insights might result from such rigorous mathematical analysis and formal theorems.

2.3 Examples of Bad RedCD

2.3.1.1 Three bearings on one shaft.
This is a classic example of overconstraint causing overload and binding. (Overconstraint is redundant constraint which does not have the benefits but does have the faults of redundant constraint.)

Once, when I was an electro-mechanical consultant to a radar company, the company's engineers showed me a servomechanism built on the same printed circuit board as its electronics. They could not adjust the mechanism to turn freely.

A wire cable had been attached with many cable clamps to hold it rigidly in place, and the mechanism shaft had similarly been attached with many bearings. (If three cable clamps are better than two, then clearly three bearings are better than two.)

I marked the redundant bearings and asked that they be removed. Instant fix. The electronic technicians who had designed and built the servoboard never really understood my explanation. I wish my consulting work had more such little triumphs.

2.3.1.2 Dovetail slides.
Dovetail machine tool slides are based on the assumption that there is large-area contact between all three pairs of sliding surfaces. In fact, none of the surfaces are perfectly flat, the spacing between the two inclined pairs is not exactly uniform, the angles are not exactly the same, and each pair of facing surfaces is not perfectly parallel. Therefore, there must be either looseness or binding at different portions of the travel. Furthermore, whatever bending of the parts is caused by the working loads adds to the shape mismatch.

Rectangular ways are now generally used instead of dovetail ways. Right angles are easier to make accurate than acute angles, and the surfaces are more accessible to precision grinding, but the basic problems remain.

To compensate these imperfections it is customary to provide a gib with a row of adjusting screws. With skilled adjustment the looseness and binding can be reduced to some minimum, but they cannot be eliminated entirely. Therefore, there is always looseness or binding, limited only by the accuracy of manufacture.

With modern machine tools it is possible to manufacture with very high accuracy, and with hand scraping it has always been possible to do so. The penalty for doing so, of course, is cost.

Heavy loads are the very good reason for large-area sliding surfaces. Elastic deflection, wearing in, and pressure lubrication tend to compensate the theoretical RedCD limitations.

2.3.1.3 Bolted feet. Bolted feet are a common form of overconstraint (Fig. 2.4). Although each foot is nominally a single point of constraint, if the foot surface is imperfectly parallel to the mounting surface, it constitutes two points of constraint, one where it touches the mounting surface and one where it touches the mounting bolt. The combination produces a moment $F \times d$, which may be harmful. Solutions to this problem will be discussed in Chap. 3.

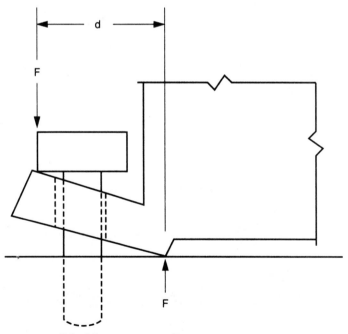

Figure 2.4 Bolted foot (exaggerated).

2.3.1.4 Lead screw.
The screw should be parallel to the ways of the driven slide, the nut axis should be coaxial with the screw axis, and the screw's bearing axes should be parallel to the screw axis (Fig. 2.5). It is impossible to manufacture these relationships perfectly; so there is either looseness or binding of the sliding and rolling surfaces and bending of the screw. A MinCD solution to this problem is given in Chap. 3 (Fig. 3.24).

It is customary, with such RedCD problems, to spend enough money on accurate manufacturing that the looseness and binding are small enough to be tolerable. The cost-saving alternative is MinCD.

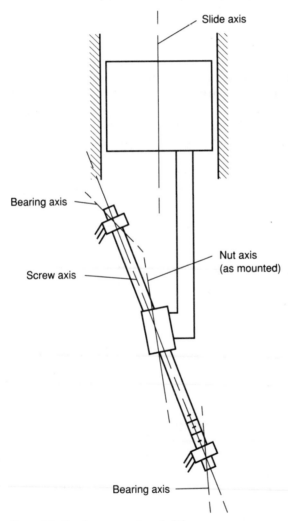

Figure 2.5 Lead screw, nut, and slide.

Degrees of Constraint 17

2.3.1.5 Chairs and tables. Chairs and tables are examples of overconstraint we meet with daily. Unless the overconstraint is very small and is deformed out by the weight load bending the table or chair, we have an annoying wobble.

I do not want to suggest that all redundant constraints are bad; some are very good indeed. The important thing is that you understand MinCD, semi-MinCD, and RedCD and learn to judge, in each case, which is the best.

2.4 Examples of Good RedCD

2.4.1.1 Cylinder head. The cylinder head on a car is a case of good redundant constraint. The cylinder head and the engine block are both rigid bodies with a flat machined interface. Two studs would hold them together in their nominal relationship, yet large numbers of studs are always provided. The reason, of course, is that the high pressure generated in the cylinders would either break the cylinder head or bend it enough to leak unless high clamping pressure were distributed around the head. Furthermore, a very large distributed pressure is necessary to compress the head gasket enough to prevent leakage. A functioning MinCD could be made by designing the cylinder head and the engine block stiff enough and strong enough, but the size and cost of the parts would be prohibitive.

2.4.1.2 Flanged joint and bolt circle. Flanged joints tied with bolt circles in tanks and pipes are a class of good redundant constraint. The flanges deform enough to take up the tolerances in the abutting surfaces.

2.5 Examples of MinCD

In this section you will see examples of MinCD in daily life and in industrial mechanisms.

2.5.1 The ubiquitous tripod

Tripods are used for cameras, surveying instruments, telescopes, tricycles, airplane landing gear, cranes, trailers, and laboratory instruments; and, of course, three-legged stools. Examples are:

2.5.1.1 Surveyor's instrument tripod. The three feet provide three vertical-axis point constraints. In addition, each foot has friction constraints in X and in Y. This simple tripod is actually overconstrained

18   Constraint Design

by the six friction forces, but in practice they don't matter because the feet slide along until they settle down with harmless small overconstraint forces in the legs. A theoretically pure MinCD tripod is shown in Fig. 2.6. One of the feet has a caster 1 and another has a wheel 2 oriented to roll toward the third foot 3. This tripod is pure six-constraint MinCD. (Figure 2.6 is intended as an exercise to demonstrate MinCD principles. I have not seen a task where it was worth actually using it.)

2.5.1.2   One-leg stool.   The one-leg stool is an amusing form of tripod in which the sitter's legs are the other two legs. There really are some

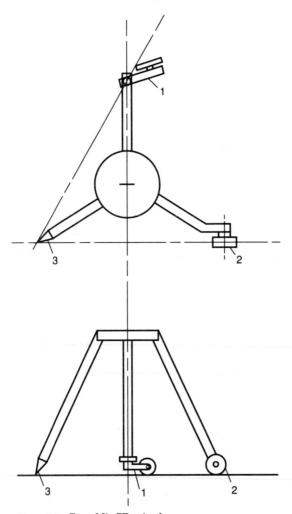

Figure 2.6   Pure MinCD tripod.

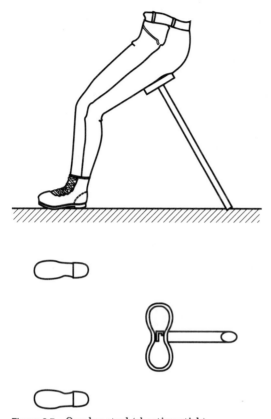

Figure 2.7 One-leg stool (shooting stick).

milking stools with one leg and a lap strap so that the milker can move the stool around without using hands. The English "shooting stick" (Fig. 2.7) has a handle which opens to form a seat; it was originally intended for the comfort of a wealthy hunter who sat on it with his shotgun on his knees while beaters drove game birds toward him.

Think, for a moment, about those human legs forming parts of these tripods. They constitute very complex servomechanisms which establish positions dictated by their brain. The subject of a human as part of a mechanism is the subject of Sec. 3.3.

2.5.1.3 Three-leg chairs. Two tripod restaurant chairs which I have encountered were instructive. In Denmark the three-leg chair of Fig. 2.8 is compact, comfortable, and stable. The forward weight of the body tends to tip the chair forward, left or right, and the human leg unconsciously braces to prevent tipping. (Remember that human

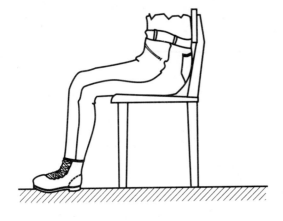

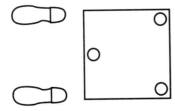

Figure 2.8 Danish chair.

servomechanism?) In England the three-leg chair of Fig. 2.9 has two legs adjacent to the sitter's legs, so the sitter's legs cannot aid stability, and a single leg in back, so one has to be careful not to tip over sideways when leaning back.

2.5.1.4 Kettles. Old-fashioned kettles often had three stubby legs so that they could be put down on a table or floor without tipping.

2.5.1.5 Ancient tripods. The ancient Romans and Greeks had very beautiful bronze braziers on bronze tripods, some of which can be seen in museums. Homer refers to tripods as articles of high value.

2.5.1.6 Tension tripods. Hanging lamps in churches are often suspended on three chains, forming a tripod with tension legs instead of compression legs. Hanging microphones in theaters are similarly suspended with their wires as one suspension leg.

2.5.1.7 Tripod derricks. Three-leg derricks are shown in Fig. 2.10;

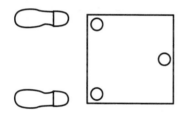

Figure 2.9 English chair.

a has three compression legs, *b* has two compression legs and one tension leg, and *c* has one compression leg and two tension legs.

When the lift axis lies within the footprint of the legs, all legs are in compression. When the lift axis is outside the footprint, either one leg is in compression and two are in tension or two are in compression and one is in tension.

2.5.1.8 Surface plates. Rectangular granite surface plates are supported on three points on a metal frame, which in turn is supported by a four-leg table. The redundantly constrained support of the table by the floor may deform the table and the frame, but no deforming strains can be transmitted to the granite block.

2.5.1.9 Machine tools. Some small precision machine tools, such as surface grinders, are built with three feet so that installation cannot deform the machine.

2.5.1.10 Jigs and fixtures. Jigs and fixtures for machining or gauging rigid parts such as castings usually clamp the parts with MinCD. The

22 Constraint Design

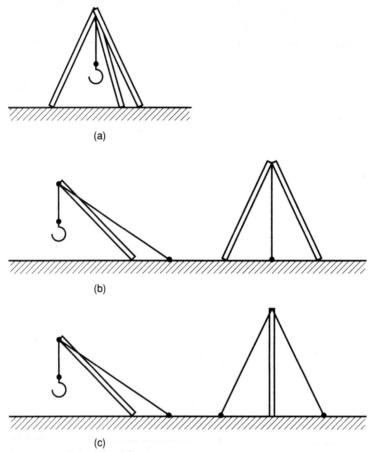

Figure 2.10 Tripod derricks.

part is seated against defined locating points without depending on operator skill and then clamped with a screw or toggle seating force.

Jig and fixture design is the only field I know of in which MinCD is customary. I think this is a result of experience, pragmatic trade practice, and intuitive understanding rather than of any theoretical insight. Several manufacturers make parts which are intended for tool designers but are also useful in other mechanism designs (see Sec. 9.11).

2.5.1.11 Tricycles. Tricycles are tripods with wheels at their feet. Small ones are used as children's toys, larger ones in motorized adult toys, larger ones in three-wheeled cars, and medium to much larger ones as aircraft landing gear. In all these cases there is a controllable yaw constraint of one wheel for steering.

Degrees of Constraint 23

2.5.1.12 Trailers. The trailer support system most commonly used is a tripod consisting of two wheels on a rear axle and a single-point trailer hitch at the tractor. Small personal trailers use a ball-and-socket hitch to prevent torque constraint at the hitch. Trucks use a set of gimbals pivoted in pitch, roll, and yaw. The yaw pivot is the disconnect and has a large pivot area which is the "fifth wheel."

2.5.1.13 Bell striker. The striker log for a Japanese bell is an example of MinCD (Fig. 2.11). Each end of the log hangs from a pair of chains arrayed in a V. (Yes, this is not really a tripod.) Gravity provides the restoring force to maintain tension in the chains and provides a soft centering constraint for this pendulum. ("Soft" and "hard" centering constraints will be described in Sec. 3.2.)

A purpose in giving these nonindustrial examples is to show the value to a design engineer of constant observation and analysis of his or her environment to develop insight. Now let's go to industry.

2.5.2 Examples of MinCD in industry

2.5.2.1 Lathe chucks. Three-jaw chucks are really examples of semi-MinCD, since we assume that the cylindrical shape of the workpiece fits the parallel gripping edges of the jaws. Four-jaw chucks are RedCD.

2.5.2.2 Robot grippers. Robot grippers may be either MinCD, semi-MinCD, or RedCD. A robot gripper is an automatic fixture and benefits from MinCD just as a jig does. Furthermore, the robot gripper must grip the part during transfer from the originating workholder to the receiving workholder. During the transfer the part may be temporarily gripped by both a workholder and the robot gripper: a severe case of overconstraint.

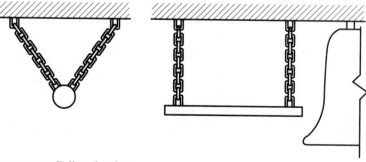

Figure 2.11 Bell striker log.

Jamming forces may be eliminated by mounting the robot gripper itself with enough freedoms so that there is no conflict (i.e., overconstraint) during these transfers. These freedoms may either be loose or be urged to a definite position by springs or otherwise. As you may imagine, my own practice is to use MinCD wherever possible, and I have designed many successful robot grippers with MinCD and self-aligning freedoms. Robot grippers are discussed in Chaps. 13 and 14, with further examples.

2.5.2.3 Straight-line mechanism. A straight-line motion I used in a series of modules for modular robots is shown in Fig. 2.12. The stationary structure consists of a steel tubular spar 1, a precision hard steel shaft 2, and two end members 3, all fastened rigidly together. The tube and shaft are nominally parallel but without great accuracy.

The moving slide consists of a box member 4 in which is welded a steel tube 5. A pair of self-aligning linear ball bearings 6 supports and guides tube 5 along precision shaft 2. Each bearing has four rows of balls on rocking, self-aligning races (Fig. 3.34). The bearings and shaft comprise a commercial matched set. Fixed ball-bearing roller 7 is mounted on member 4 and rolls on one side of tube 1; spring-loaded roller 8 rolls on the other side of tube 1.

Bearings 6 and shaft 1 constrain slide 4 in X, Y, pitch, and yaw. Rollers 7 and 8 constrain in roll. The Z constraint is the motor or air cylinder which generates Z motion to do the work of the module.

Tolerance in the diameter of tube 1 is taken up by the seating force of roller 8 and its spring. Tolerance in the parallelism of tube 1 and

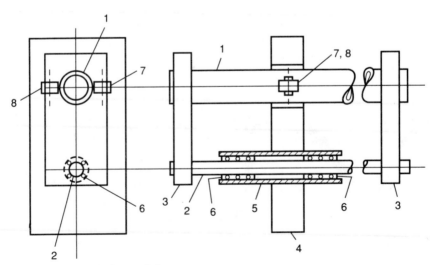

Figure 2.12 Small robot module.

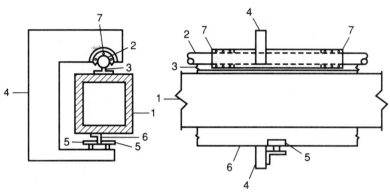

Figure 2.13 Intermediate robot module.

shaft 2 can produce no binding; insofar as they are skewed, the path of the slide assembly has a small helix. (For most robots the requirement for path shape is repeatability rather than accuracy; see Chap. 13.) *Thus this is an inexpensive mechanism with zero looseness and zero binding.*

2.5.2.4 Larger straight-line mechanism. A larger linear motion for several middle sizes is the structure of Fig. 2.13. Spar 1 is an aluminum extrusion from 4 in square to 8 in square. Commercial precision shaft 2 on its stand-offs 3 provides the counterpart of shaft 2 in Fig. 2.12. Slide 4 carries two matching commercial self-aligning linear ball bearings 7, which with shaft 2 comprise a matched set. A pair of rollers 5 engages rail 6 for roll constraint. These middle sizes were made up to 20 ft long and were used both vertically and horizontally. Again, the cost was low; very little machining was required, none of it over large areas. For better straightness than provided by the raw extrusion, we shimmed the stand-offs 3 by using the simple instrumentation of Fig. 3.23. A variety of drives, mountings, and load attachments were used, all independent of this basic five-constraint, one-freedom MinCD mechanism.

2.5.2.5 Large linear-motion mechanism. The largest linear-motion module we made was 140 ft long, in three sections, and carried an 800-lb payload. It is shown in Fig. 2.14. It was MinCD, pure. The spar is a ½-in steel plate 1 welded to the flanges of a 24-in-deep I beam 2. The enclosed box section provides torsional stiffness. (In Sec. 2.6 is explained how we fixtured and welded the combination with no warping and without big equipment.) Carriage 6 rolls on the edges and sides of the plate on six wheels 3, 4, and 5, providing five constraints and one

26   Constraint Design

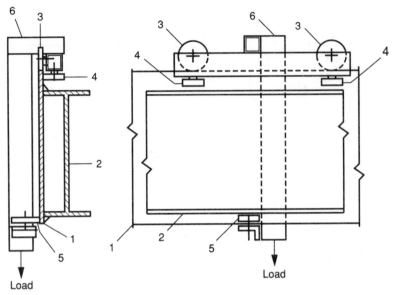

Figure 2.14 Large robot module.

freedom, with gravity as the seating force. The spar is carried on short cantilever I beams from four building columns, and a pure MinCD bracket attached the spar to the cantilevers with adjustability to line up the sections.

Very large cranes of this general configuration have been made to carry ship containers.

2.5.2.6 Large storage and retrieval robot. A large MinCD machine which we made for automatic storage and retrieval robots is shown in Fig. 2.15. It uses a vertical module of the type shown in Fig. 2.13, up to 20 ft high, and moves it horizontally up to 150 ft. The weight of the vertical spar 1 is carried by a single large wheel 2, which has a solid polyurethane tire for quiet running. The lower end is guided by matched-set wheel pairs 3 and 4 rolling on steering rail 8, and the upper end is guided by similar wheel pair 6 rolling on steering rail 9. Pinions 5, 7 at the ends of shaft 10 engage racks 11 and 12. The six constraints 2, 3, 4, 6, 5, 7 fully constrain spar 1 and do so without redundancy. Shaft 10 is turned by motor 13 to move the vertical spar along the horizontal track. Adjustable coupling 14 permits relative rotation of the pinions during setup to adjust the vertical spar into true vertical. Coupling 14 is shown in greater detail in Fig. 3.21.

To be absolutely pure, each of the wheel pairs 3, 4, 6 should have one fixed roller and one spring-loaded roller. However, the steering

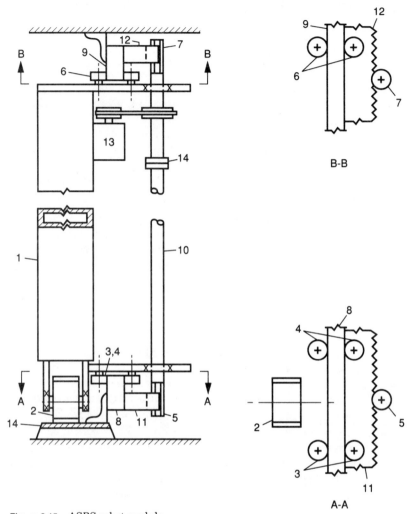

Figure 2.15 ASRS robot module.

rails 8, 9 varied no more than a few thousandths in thickness; so, for simplicity, we used matched sets comprising cam followers with eccentric adjustments and set up fixed spacing as in Fig. 2.3. We ignored the backlash between the pinions and the racks, although we could have added antibacklash gears, with springs for tooth-seating forces, if zero looseness in this direction were important (see Chap. 16, "Backlash").

The important things about these MinCD examples is that we could make very large mechanisms with no looseness and no binding and do so inexpensively, with very little machining, using only small machine tools, and with only moderate assembly skill. In fact, our shop was shorter than the longest machines that we made, so we had to ship them without a complete trial assembly; yet they went together perfectly in the field.

These machines were made without machining large areas or large parts other than drilling holes. The straightness of their motion paths was limited only by the skill of human adjusters and, for our market, was never required to be closer than $1/32$ in from a straight line. For robot service this is much better than necessary because *repeatability* rather than *accuracy* is what is needed. Repeatability, in these large MinCD machines, is actually within a few thousandths of an inch. If we had wanted a narrower tolerance, we could have had it with more careful adjustment. (See Chap. 13 for robot theory.)

A note about accuracy: mechanisms are made with accuracies and precisions from a microinch to a foot. The principles in this book apply to the entire spectrum; in fact, most instruments in the extremely close precision range tend to be made MinCD. Do not be misled by the accuracies named above, which are coarse by machine tool standards, suggesting that the principles of MinCD apply only to mechanisms of low accuracy.

2.5.2.7 Safety caging. Safety caging is a set of constraints which influence the position of a part or assembly only if there is a failure of the working constraints or controls. The safety chain on a trailer hitch is a simple example; it comes into action only if the working constraints, the ball coupling, fails. Overtravel shock absorbers are safety caging soft constraints which come into action if the working controls fail and if the overtravel limit-switch controls also fail. In rail-guided systems, hooks are sometimes provided which normally clear the rails by a substantial margin but which limit the extent of a wreck if there should be failure of the normal guidance configuration.

2.5.2.8 Assembly. Figure 2.16 shows two parts 1, 2 which must be assembled easily with screws and nuts but with an absolute minimum of variation from assembly to assembly. Three screws and nuts (not shown) are used to join the parts; they provide a tripod assembly in which washers 3 prevent rocking due to uneven surfaces. Holes $1A$ and $2A$ fit closely to the screw diameter and provide two axes of relative constraint. Hole $2B$ and the *width* of slot $1B$ fit closely to the screw diameter and provide a third relative constraint. *These constraints are independent of the tolerance on hole spacing, which is usually greater than the tolerance on hole diameter or slot width.* Hole $2C$

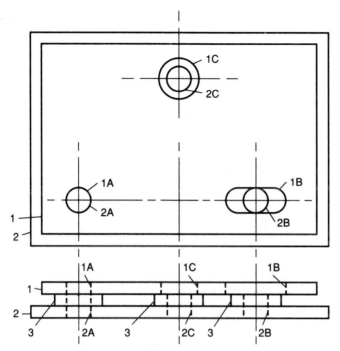

Figure 2.16 Two-part assembly.

fits closely to the screw diameter to give a definite position to its screw, but hole 1C is large enough to clear the screw regardless of hole-to-hole tolerances. Screw tension provides the seating force. Thus the parts are easily and accurately joined with no looseness other than the small tolerances on hole and screw diameters.

2.5.2.9 Tandem shafts. A common problem is to support a very long shaft with a plurality of spaced bearings to prevent whip and excessive sag. To do so is RedCD and requires very careful alignment and a very rigid support structure. Figure 2.17 shows a MinCD solution to the problem. The shaft is cut into sections joined by universal joints 2.

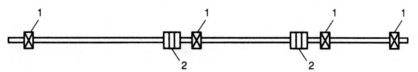

Figure 2.17 Chain of shafts.

A bearing 1 is placed adjacent to each U joint. Only one end section is on two bearings.

2.6 MinCD with Flexible Bodies

So far this book has implied, without saying so, that you deal only with rigid bodies; the theory of MinCD as presented so far has implied this assumption. Not all the bodies you deal with in the real world are rigid, and the flexibility of these bodies affects the ways in which they should be constrained.

Although all materials are deformable and therefore all bodies are somewhat flexible, many bodies can be approximated as rigid for engineering purposes. Examples are cast or welded machine tool frames, parts machined out of a solid, and sheet-metal boxes with internal bracing.

What is MinCD for flexible bodies?

A flexible body has minimum constraint if the actual constraints, no matter how many, do not impose forces other than those needed to restore or prevent the body's flexibility deflections in addition to positioning the body as if it were rigid.

2.6.1 Examples of flexible body

2.6.1.1 Long machine beds.
A long machine tool or printing-press bed may have a deflection tolerance of only a thousandth of an inch under its own weight plus a moving load of several thousand pounds in the machine tool. Therefore, it may need a multiplicity of jacks, shims, or wedges along its length to support it against gravity and against working loads in its unstressed straight form. This is not redundant constraint since each support does its own work without interfering with the work of the others, and the only part deforming done is to *remove and prevent* the deformation done by gravity and working loads. Such supporting constraints added to the minimum constraints positioning the body are called "forming constraints."

After a flexible machine bed or rail has been MinCD-positioned and multiple-point-supported, it is often the practice to add grout under the bed or rail. This grout then supplements or replaces the adjusting wedges or jacks. (Grout is a hardening matrix such as portland cement, polyurethane, silicone RTV, asphalt, or epoxy. It may have fillers such as gravel, sand, metal particles, cork, or miscellaneous fibers.)

Grout is also used to lock a part into position after it has been adjusted. An example is placing a pair of sleeve bearings in loose holes, aligning them by a shaft through both, and then locking them into position with a plastic cement. This eliminates line-boring the bearings after assembly.

A form of grout used in electronics is potting compound. It is a hardening plastic in which an electronic assembly is embedded for support and protection. In a sense, potting is the ultimate in RedCD.

An unusual form of grout is water poured into a honeycomb part on a milling machine and then frozen. The ice supports the webs of the honeycomb while they and the ice are being milled to shape.

2.6.1.2 Robot spar. The large robot spar of Fig. 2.14 is a ½-in by 30-in by 45-ft steel plate welded to the flanges of a 24-in-deep I beam without warping. To constrain the parts we put on the floor a rectangular field of jackscrews on 1-ft centers, optically leveled their top ends, laid the flexible plate on them, and laid the flexible beam on it. After welding, the assembly was flat, straight, and rigid. There had been no redundant constraint but a great many forming constraints.

2.6.1.3 Flexible cart. A large, flexible part is positioned, or caged, as illustrated in Fig. 2.18. The "part" is actually a rolling cart 30 ft long carrying an aircraft inspection fixture. A sufficiently rigid cart does not fit in the allowed space, so we allow cart 1 to be flexible and support it by many redundant casters 9; it flexes slightly as it rolls over the floor. When it is in approximately the desired position in inspection machine 8, air cylinders extend pointed pins 3, 4 through close-fitting hole 3' and close-fitting slot 4'. The pins establish two linear and one rotary constraints, preventing horizontal motion. Then a multiplicity of air cylinders 6 lifts a set of cart lugs 5 against a corresponding set of fixed stops 7 in the inspection machine. This lifting action raises the casters off the ground and forces the cart to conform to the

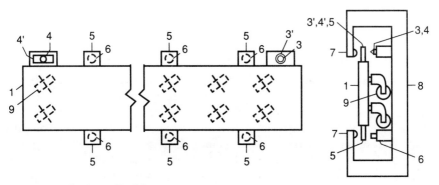

Figure 2.18 Caging a flexible cart.

level plane established by the fixed stops. *This example also introduces the concept of "sequential caging"; the time sequence in which constraints are applied can be useful and important.*

This example shows that the concept of minimum constraint, while requiring only six constraint points for a rigid body, can incorporate additional constraint points to restore a flexible body to its desired shape.

It can be useful to make part of an otherwise rigid body flexible to permit it to self-align with a desirable redundant constraint.

2.6.2 Classes of flexible body

2.6.2.1 Thin. Thin bodies may be flexible in one direction of bending and in twist and rigid in other directions. An example is an airplane wing.

2.6.2.2 Long. Long bodies may be flexible in two directions of bending and in twist. An example is a railroad rail.

2.6.2.3 Large. Large bodies which seem to be rigid may actually deflect under weight and load more than their requirements permit. The bed of a long milling machine is an example; a deflection of only a few thousandths of an inch under a moving load of several thousand pounds may not be tolerable.

Another consideration in MinCD is the stresses and deflections of both body and locators near the points of contact. We have discussed the use of spherical contact members, and in a later chapter we will consider semi-MinCD. Adding a multiplicity of constraints which do not exert forces on each other but do distribute the load between the constrained body and its locators is consistent with the principles of MinCD.

2.7 Load Dividers

Whiffletrees are used to distribute a constraint force among a plurality of points. *"Whiffletrees?"* The word originated in agriculture and refers to linkage arrays which equalize the draft load among a number of horses pulling a wagon or plow. It appeared in the aircraft industry to describe a three-dimensional linkage array leading from a hydraulic cylinder to many rubber pads cemented to the surface of a wing to duplicate, for testing, the distribution of aerodynamic forces on the surface (Fig. 2.19). Dimensional tolerances of the levers and

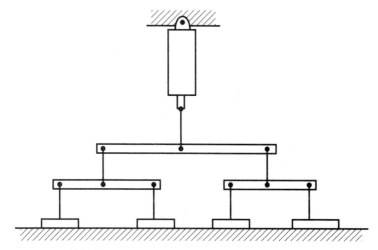

Figure 2.19 Whiffletree.

tension tie rods have no effect on load distribution; the levers simply rotate to accommodate them.

An inverted array will conduct the weight of a fragile load from a plurality of support constraints down to a single support point. The linkage equalizes the force on all the support constraints if the links are symmetrical or, as in testing those airplane wings, distributes the force in any way you wish.

An example of this whiffletree system is in oriental architecture, in which tiers of "brackets" are used in this manner to spread the supporting force of columns over a large area of roof (Fig. 2.20). Actual pivots are not used; the links simply rest on each other.

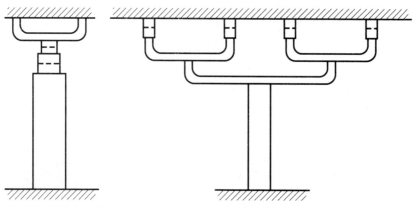

Figure 2.20 Oriental roof brackets.

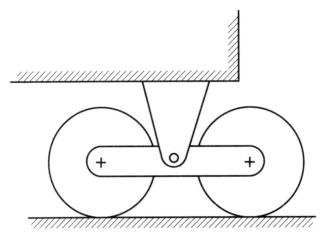

Figure 2.21 Truck axles.

The whiffletree linkage is quite old. I have seen an example in Beijing in which the weight of the imperial sedan chair was divided among a large number of bearers, simultaneously smoothing the ride. At the opposite end of the technology spectrum, each mirror in the new Keck giant telescope is supported at 12 points on a whiffletree to reduce gravity sag.

The art form called a mobile is a whiffletree.

Another load-distributing array is our old friend the tripod, which divides a load among three support points. You could make a whiffletree of tripods (triffletree?).

The distribution of a gravity load among a plurality of train or truck wheels is conventional. The distribution is done by a linkage leading from the load to the axles. Figure 2.21 shows a simple two-axle load-dividing link, but more elaborate linkages are used in heavy trucks to divide the load among three or more axles.

The essence of minimum constraint design is not in limiting the number of constraints to the minimum number required for a rigid body but rather in *eliminating overconstraints*.

Chapter 3

Kinds of Constraint

3.1 Hard Constraints

Hard constraints occur at the contact point between rigid bodies. (I know that no body is perfectly rigid. For our purposes a good approximation is to consider that steel is rigid and rubber is soft.)

3.1.1 Examples of hard constraints

3.1.1.1 Point and surface. The point-and-surface constraint is exemplified by the instrument tripod. It is simple and inexpensive. The point contact stresses can be quite high and cause elastic and plastic deformation to both foot and ground. (Sometimes this may be a benefit if the points dig in and anchor the tripod on the side of a hill.)

Point-and-surface constraint can be a disadvantage when the friction between foot and ground causes tangential as well as normal force and results in severe overconstraint. Figure 2.6 shows a combination which is pure because only one foot of the tripod is a point with two friction constraints, the other feet having one and none, respectively.

3.1.1.2 Ball and surface. A ball-and-surface constraint has a well-defined contact point but has much lower stresses in the parts than a point-and-surface constraint. Sliding into equilibrium position during setup is much easier than with point and surface. The stresses are Hertzian stresses, and their computation can be found in textbooks and handbooks. The contact between a ball-bearing ball and race is the most common place where such contact exists. The constraints of Fig. 2.1 are ball and surface.

3.1.1.3 Roller and surface. A roller-and-surface constraint has less contact stress than a ball-and-surface constraint because there is line contact *if* the roller and surface are self-aligning (Fig. 3.28) or are parts of a matched set, as in a roller bearing.

3.1.1.4 Shaft-and-sleeve bearing. A shaft-and-sleeve bearing has lower contact stress than a roller-and-surface constraint. It provides two linear constraints; if the length of fit is sufficient or if there is a matched set with one shaft and two bearings, it also provides two rotary constraints. Typically the contact stress is low enough that there can be relative sliding in operation, which is what a sleeve bearing is all about. The mounting of the bearings can be MinCD, semi-MinCD, or RedCD.

3.1.1.5 Ball and socket. A ball-and-socket constraint has low stress, provides three linear constraints instead of two, and has no angular constraint. It is used in many devices, as we shall see.

3.1.1.6 Bolted foot. A bolted foot is intended to provide a small area constraint with an integral seating force (the bolt tension) and a pair of lateral friction constraints. We have seen that it can also introduce damaging moments unless spherical-washer pairs are sandwiched under the foot and preferably under the bolt head as well (Fig. 3.42). Internal stresses in the part from opposing friction forces among the feet can be minimized, with installation skill, by tightening the bolts cyclically, a little at a time, as one tightens the nuts holding down a cylinder head.

3.1.2 Examples of wheel constraints

3.1.2.1 Single narrow wheel. A single narrow wheel rolling on a smooth surface provides a point of hard constraint where it touches the surface. The body carrying its axle has two linear constraints, up and down and side to side, and one linear freedom, the direction of rolling. If the wheel tread is narrow, the wheel can rotate in yaw about the small area of contact with the surface. An example is a wheelbarrow.

3.1.2.2 Pair of wheels tight on a common axle. A pair of wheels tight on a common axle or a single wide wheel tread differs from a wheel with a single narrow tread in that it provides a roll constraint and a yaw constraint in addition to two linear constraints.

3.1.2.3 Pair of wheels loose on a common axis. A pair of wheels, each loose on a common axis, permits a yaw freedom in addition to the pitch and forward freedoms.

3.1.3 Examples of wheels on tracks

Wheels are constrained to follow straight and curved tracks in several ways:

3.1.3.1 Two flanges on one wheel. A wheel with two flanges will follow a track, but with looseness equal to the difference between flange spacing and track width. Furthermore, the friction between flange and track can be noisy and eroding.

3.1.3.2 One flange on each wheel of a pair. A pair of wheels with one flange on each wheel on a double track behaves in the same way as two flanges on one wheel.

3.1.3.3 Traditional railroad wheels. Railroad wheels are slightly conical, and both wheels on an axle are fixed to the axle with the apexes of the cones outboard. If the wheels drift off the centerline of the track, one wheel rolls on a smaller diameter and the other rolls on a larger diameter. The axle then rolls in a circle in such a direction that it steers back to the centerline of the track. Track following is usually stable, but in some combinations of multiple axles and a heavy train there can be oscillation.

3.1.3.4 V grooves. V-shaped grooves on some wheels roll on corresponding V-shaped tracks. A common track is commercial angle iron laid on the ground with its hypotenuse down.

Some wheels have V-shaped edges and roll in matching grooves in tracks. Some very accurate commercial systems of this sort are made. (See Sec. 7.1.)

3.1.4 Examples of rotary hard constraints

3.1.4.1 Jaw clutch. A jaw clutch is an angular hard constraint which has a discrete number of angular positions in which it can be engaged.

3.1.4.2 Splined shaft. A splined shaft is an angular hard constraint having a linear freedom along the axis of the shaft. Some splined shafts are made with ball bearings on the linear freedom.

38   Constraint Design

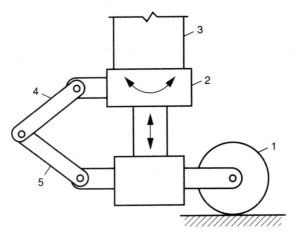

Figure 3.1   Rotary constraint.

3.1.4.3 U joint, splined shaft, U joint.   The combination of a U joint, a splined shaft, and a second U joint provides a single angular hard constraint with three linear and two angular freedoms (Fig. 2.2).

3.1.4.4 Independent rotary constraint.   A rotary constraint independent of a linear freedom on its axis is illustrated in Fig. 3.1. This is the steering nose-wheel suspension used in aircraft. Wheel 1 is steered by mechanism 2 but is free to rise and fall in cylindrical shock absorber 3. Links 4 and 5 constrain rotation in the shock absorber without interfering with its linear motion. The links are wide; so there is no twisting in their joints.

Hard constraints are sometimes made adjustable or variable as described in Secs. 3.6 and 3.7.

3.2 Centering Constraints

3.2.1 Hard centering

Hard centering constraints locate a moving part until the disturbing force or torque exceeds a preset value and then yields (Fig. 3.2). Typically the preset force is established by suppressed springs, as in the figure. Figure 3.2a shows two springs, not necessarily of equal force, for the two directions. Figure 3.2b shows a single spring operating in both directions and assuring that the centering force is exactly equal in both directions.

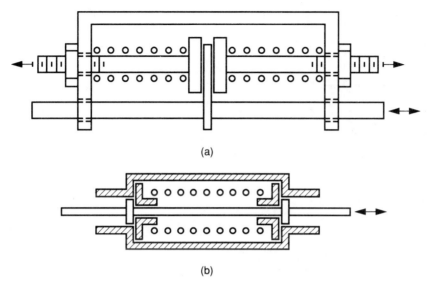

Figure 3.2 Bungee. (a) Two springs for the two directions. (b) One spring for both directions.

The term "bungee" is sometimes used for centering springs although that word is used differently in aircraft.

Centering constraints can be made for one or more axes of motion and can be symmetrically or unsymmetrically loaded, as in Fig. 3.2. The illustration shows centering constraints for linear motion. It should be obvious that analogous centering constraints can be made for rotary motion.

Figure 3.3 shows a joystick bungee with pitch, roll, and yaw bidirectional spring constraint and down unidirectional spring constraint, all from one spring.

Pneumatic or hydraulic cylinders can be used in place of springs in hard centering constraints.

3.2.2 Soft centering

Soft centering constraints urge a movable part toward a central position, but the part may be displaced by any force, the amount of displacement being a function of the amount of force. A gravity pendulum has such a soft centering constraint. A part on the end of a spring is another example. A watch balance wheel has rotary soft centering from its hairspring. A sensitive torsion instrument hung on a quartz fiber has rotary soft centering. A floating object has vertical soft cen-

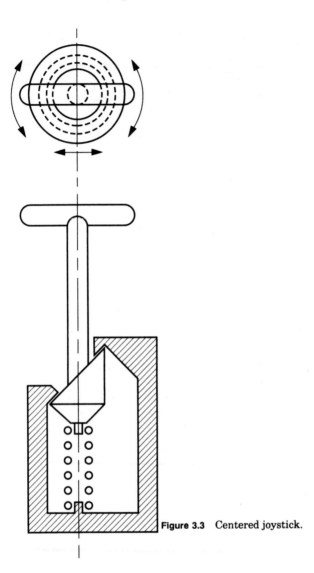

Figure 3.3 Centered joystick.

tering from its buoyancy. Buoyancy as a mechanism element is discussed in Sec. 3.4.

3.3 Human Constraints

Human beings are an integral part of many stationary structures and many operating mechanisms. Humans drive cars, fly airplanes, and control machines and tools. In doing so they apply both stationary constraints, as in aiming a rifle, and dynamic constraints, as in steering a car or riding a bicycle or skateboard.

Figure 3.4 shows a simple device used to carry a crucible in a foundry in which two humans provide the legs of a tripod with their arms and hands. One person carries handle 1, and a second person carries handles 2 and 3. Only the second person controls the pitch axis to pour from the crucible; so there is no problem of accurate cooperation.

A wheelbarrow is an example in which a person provides two legs of a tripod and controls all axes of motion.

Sometimes human constraints can be described geometrically as when the mechanical axes of the constrained device establish the axes of constraint as in an airplane joystick; sometimes they are rather more complex, as in carrying a squirming baby.

Humans are almost always RedCD because the human has a multiplicity of constraining muscles and surfaces. Pushing a button is the only exception I can think of.

In most cases human constraints are part of an enormously complex feedback control system. Sight, sound, and touch are only some of the feedback sensors; the brain is the control computer, and several thousand muscles are the actuators.

A wine connoisseur uses smell and taste as feedback sensors in managing the wineglass. Handicapped people use their available muscles to operate prosthetics. You will remember, from Chap. 2, the monopod milking stool and shooting stick and the Danish three-leg restaurant chair.

In some cases the human can use inertia as the seating force to stabilize an otherwise unstable structure. Examples of such structures

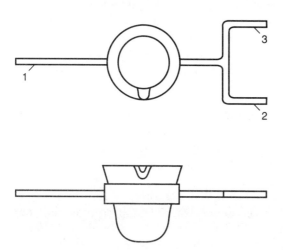

Figure 3.4 Three-handle foundry ladle.

having only two points of hard constraint are a person on a bicycle or scooter, a wire walker (with and without a balance bar), a person on stilts, and a ballerina on two toes. Examples of such structures having only one point of hard constraint are a person on a unicycle, a person on a pogo stick, and a ballerina on one toe. (An automatic servo-controlled pogo stick has been built and made to work, and such a device was once seriously proposed as a moon surface vehicle.)

If you design machines for human use and control, it behooves you to learn something of human engineering ("human factors," "ergonomics"). Reference 40 in Chap. 20 is a major sourcebook on the subject, and there are many shorter ones. The easiest source is observing your own behavior.

3.4 Soft Constraints

So far we have considered constraints to be contacts between parts made of hard materials so that the constrained positions are accurately established. In this section we will consider constraints in which at least one part is not hard. A small relative displacement causes only a small incremental force. It will sometimes be uncertain as to whether a soft element is a constraint or a seating force, but it usually does not matter.

3.4.1 Uses for soft constraints

3.4.1.1 Shock and vibration isolation. Solid-material shock mounts are used as interfaces between a shock- and vibration-sensitive assembly and a supporting structure which has vibration and shocks. A typical example of a need for such isolation is an electronics cabinet in an airplane or tank. Shock mounts are also used to support vibration *sources*, such as automobile engines, to attenuate vibration transmitted to the underlying structure. Shock mounts are used to support machines which generate large shocks and vibrations to prevent these disturbances from being transmitted to the ground. Cork pads are often used for this purpose. Drop hammers and heavy punch presses are examples of such machines. Soft-constraint supports are discussed under the headings "Shock Mounts," "Buoyancy," and "Magnetic Attraction and Repulsion" in Subsecs. 3.4.4 and 3.4.5.

3.4.1.2 Oscillation damping. Oscillation damping is the damping of oscillation generated within the device itself. Damping materials are applied to surfaces having oscillating strain. Weights are coupled to parts of the device via damping materials. Acoustical noise energy is dissipated by absorbent surfaces.

3.4.1.3 Contact stress reduction. Elastomeric feet of many kinds are used as load spreaders. The rubber tire on wheels, both solid and air-filled, is most common. The elastomer spreads its load over a footprint and thus reduces stress. It reduces rolling noise, and it acts as a shock mount. Polyurethane is a strong and creep-resistant elastomer now used for solid tires. (See Fig. 2.15.)

3.4.1.4 Scratch and dent prevention. Felt, cork, and cork and elastomer mixtures protect surfaces and block sound transmission. (Cork also has a particularly high coefficient of friction. I once was able to save a design by using cork against cork in a clutch which did not have to engage when running. It was able to transmit more torque in a limited size than we had been able to do before.)

3.4.1.5 Pressure distribution. A trapped rubber block acts rather like a bag of incompressible oil and exerts almost uniform pressure in all directions. Rubber pads spread the force of a hydraulic-press ram over a multiplicity of shearing and forming dies.

Many years ago I invented a zero-insertion-force, multicontact electrical connector by using this principle (Fig. 3.5). Rows of contacts 1 are made as printed circuit areas. One set is made on thick, rigid plastic 2, and the other set is made on thin, flexible plastic 3. Each thin set is laminated to soft rubber cores 4. Both sets are spaced on soft springs at sufficient distance that they can be interleaved without touching; so there is substantially zero insertion force. After the sets are interleaved, a lever-operated cam 5 squeezes the contacts together through a stiff spring 6. The rubber compresses at thick spots and bulges at thin spots, spreading the force uniformly over all contacts. A development model had 100 contacts, exerted 7 *pounds* of force on each contact, and was engaged and disengaged, and force applied and released, with fingertips. Measured electrical resistance was equal to the calculated resistance of the metal alone, with zero contact resistance. (I licensed the patents to a major connector manufacturer, who paid me a minimum royalty for many years while neglecting the product. But this is not the place to bemoan the obstacles between technical success and commercial success.)

3.4.1.6 Overtravel cushioning. One technique of constraining accidental overtravel of a member, by absorbing its kinetic energy over a distance, is to cause it to cut or inelastically deform solid material. One system uses a lathe tool on the moving member. It cuts a heavy chip off a block of steel in its path. Another uses a block of aluminum honeycomb which is crushed by the impact. Both are one-shot techniques

44 Constraint Design

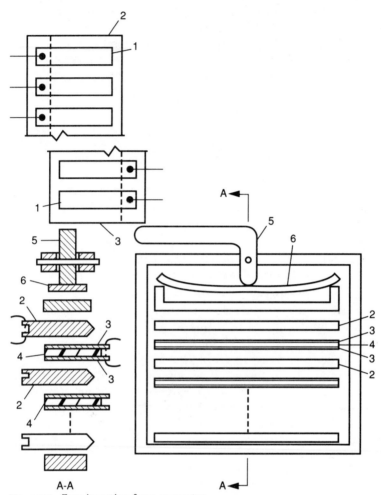

Figure 3.5 Zero-insertion-force connector.

which require replacement after the accident, but both require no maintenance and are highly reliable. Early moon exploration packages (Project Ranger) used a balsa wood sphere enclosing the instrument package to provide a cushioned landing for its contents.

Dashpots are used as overtravel cushions (Subsec. 3.4.5).

Where rebound is acceptable, a simple rubber bumper is used.

3.4.1.7 Separation of sliding parts. In linear and rotary sliding bearings, lubricating-fluid films are forced between the parts to prevent abrasion. Both liquid and gas films are used. Fluid is pumped into the space between the hard parts either by viscous drag (hydrodynamic bearings) or by an external pump (hydrostatic bearings).

The largest thrust bearings in the world are the Kingsbury type, in which a number of tilting segments generate wedge-shaped hydrodynamic films under the segments. They are used in ship propulsion thrust bearings and in the bearings which support the weight of large hydroelectric turbine generator rotors.

Dramatic examples of hydrostatic bearings are those which guide the rotation of the 200-in telescope in the Mount Palomar observatory (oil), those which lubricate granite sliding on granite in high-precision machine tools (air), and the bearings between a pallet and the factory floor (air) which enables one person to push around enormous loads.

3.4.2 Seating forces

Although I have distinguished between constraints and seating forces, this is an appropriate place to describe the elements and devices used for seating forces since they differ only in application from soft constraints.

Since most of them have been described above or are self-evident if only named, the following is a list of such sources of seating force. If a seating force is not to be a source of additional positioning constraint, it must be free to move in directions transverse to its force axis so that the positioning constraints are not opposed by the seating-force element.

- Gravity
- Linear and torsional springs
- Magnetism
- Fluid pressure
- Fasteners (bolts, rivets, etc.)

A tight fastener constitutes a stiff spring (e.g., a stretched bolt); otherwise looseness would result from any minute dimensional change. There are places where a threaded tie rod is deliberately made long to increase its spring energy when it is tightened up, and there are places where a bolt is machined down to the root diameter of its thread to reduce its spring constant without reducing its strength.

In general, fasteners which do not have springs included in the fastening stack will impose friction torques because of the tightening process in addition to the seating force. (To provide pure MinCD a fastener should only compress a spring which provides the actual seating force.) There is a system for tightening a bolt without generating torques or forces other than axial tension. A hydraulic cylinder and a second nut pull the bolt to the desired tension, after which the nut is made fingertight. When the hydraulic cylinder is removed, the nut holds the bolt at the tension established by the cylinder.

3.4.3 Materials used in soft constraints

How these materials are used is described later in the chapter.

- Elastomers
- Cork; cork and elastomer mixtures
- Leather
- Textiles (felt, cordage, woven and knit fabrics)
- Wire (knitted, rope, springs)
- Permanent magnets
- Electric currents in coils
- Eddy currents in metals
- Exposed fluids (air, oil, water)
- Enclosed fluids (air, oil)

3.4.4 Effects used in soft constraints

3.4.4.1 Elasticity. Materials are used for soft constraints because they have low elastic constants and have substantial hysteresis. Among these materials are elastomers, cork, leather, textiles, knitted wire, and wire rope in bending.

3.4.4.2 Hysteresis. Mechanical hysteresis is the energy loss when a material is subject to a varying stress. The energy loss appears as a smaller force in the return direction than the applied force in the forward direction. This energy loss is used to damp oscillation in shock and vibration absorbers. Most metals have low hysteresis. Most of the materials listed above in Subsec. 3.4.3 have high hysteresis, which is why they are used for damping.

3.4.4.3 Viscosity. Viscous force is used in various devices to generate soft constraint either to damp oscillation or to decelerate a body. Liquids whose viscous drag is proportional to speed behave with mathematical linearity (like eddy currents) and are called Newtonian. Floated gyroscopes rely on such linearity and typically use Fluorolube* oils because of their linearity. (See Subsec. 3.4.4.4, "Buoyancy.") Viscosity of all fluids is temperature-sensitive (although silicones are less temperature-sensitive than most); so such gyroscopes are typically heated to a fixed temperature.

Liquids are available with viscosities which range from less than that of water up to what seem like solids. The toy Silly Putty† is a silicone liquid of very high viscosity.

There is a phenomenon called "thixotropy" which is the liquid equivalent of stiction. Thixotropic fluids act like solids until they break loose, flow like liquids, and then revert to a solid state when flow stops. You may find use for them in a new kind of constraint.

3.4.4.4 Buoyancy. Buoyancy exerts a soft force tending to cancel gravity. If the body is completely immersed, it cancels any other acceleration of the fluid's container.

Buoyancy exerts a soft upward constraint on a boat. Buoyancy supports a pontoon bridge; so a structural bridge is not needed.

The mariner's compass is filled with liquid. Buoyancy reduces the vertical force exerted by the compass card and magnet on their bearing and thus reduces friction. The liquid also damps oscillation of the card and suppresses corrosion in this hostile sea-air environment; so the buoyant and damping liquid is a two-way winner, always an objective in design, games, and life.

A most elegant example of buoyancy constraint is the HIG (hermetic integrating gyroscope), Fig. 3.6. The gyroscope 1 is enclosed in cylindrical container 2, which fits loosely inside cylindrical container 3. The space between is filled with liquid 4. The weight and volume of container 2 are made neutrally buoyant in liquid 4. Minute deviation from neutral buoyancy is carried by sleeve bearings 5. (Since the bearing load is extremely small, the bearing shafts are merely fine wires turning in jewel bearings; the friction is much smaller than could be obtained with the best ball bearings.) Transducers 6, 7 measure the gyro angle and apply control torque. The liquid is the Fluorolube mentioned above, which provides Newtonian damping; so the behavior of the gyro is mathematically linear.

Let us not ignore the domestic water bed as an example of a most pleasant buoyancy device.

I speculate that buoyancy could be used to partially support very heavy moving machine parts such as rotating telescope domes, swing and lift bridges, and very large machine tool slides, thus reducing the size and cost of their support bearings and the friction load in driving them. Hydrostatic lift bearings can be thought of as artificial buoyancy.

---

*Fluorolube is a trademark of Hooker Chemical.

†Silly Putty is a trademark of Binney & Smith, Inc.

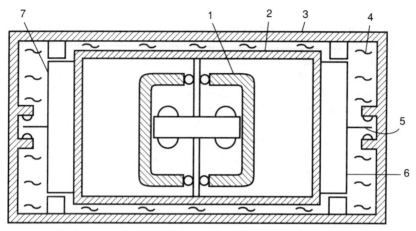

Figure 3.6 HIG gyro.

3.4.4.5 Eddy currents. An aluminum or copper member on a moving part is moved between the poles of a permanent magnet on a stationary part. Eddy currents are induced in the moving member and dissipate energy. The damping is exactly proportional to speed, i.e., is mathematically linear; there are zero forces other than speed damping; and there is zero stiction. Typical uses are in watthour meters and some old-fashioned laboratory analytic balances. Eddy-current clutches and brakes use an electromagnet instead of a permanent magnet so that the magnitude of the eddy currents can be controlled.

3.4.4.6 Magnetic attraction and repulsion. Magnetic force between permanent magnets, between current-carrying coils and magnets, or between coils and coils is used as a form of buoyancy to neutralize all or part of gravity force on a body, usually to reduce friction. The watthour meter uses pairs of permanent magnets for this purpose. Small electro-magnetic voice coils drive loudspeaker diaphragms, and very large voice coils apply thousands of pounds of force in testing shakers.

Very large soft forces can be generated by electromagnets with feedback control. Large repulsion systems use eddy currents in one part induced by electromagnets in the other part. The extreme case is the electro-magnetic–levitation railroad train.

3.4.4.7 Gravity. Most uses of gravity are as seating forces. The gravity force on a weight can be transmitted by ropes or levers for any other purpose. Gravity is inelastic; i.e., the force is constant and independent of displacement. Vertical acceleration force adds to or subtracts from gravity force.

3.4.4.8 Fluid pressure. Gas and liquid pressure, in a variety of resilient enclosures, is used in a large number of devices.

3.4.5 Devices used for soft constraint

3.4.5.1 Dashpots. A dashpot produces a viscous force which resists a time-varying position. It contains an enclosed viscous fluid, which is driven through an orifice by the motion of a piston. It dissipates energy. Thus it limits resonant vibration due either to a steady exciting frequency or to a shock. Typically it comprises a cylinder, a piston, and an orifice connected so that the fluid flows through the orifice from one side of the piston to the other.

Since a dashpot seeks no unique position within its working range, it is not a positioning constraint and may be added to a MinCD system without changing the purity of the MinCD. (But be sure that the dashpot is mounted in such a way that it imposes no constraint other than along its operating axis.) Each shock absorber in your car is a linear dashpot, and the door closer on your office door contains a dashpot. Many impact cushion devices combine a spring, a hard stop, and a dashpot.

One limitation of dashpots is the stiction between their piston rods and their seals. There is a pneumatic dashpot trade named Airpot* which reduces stiction to a very low value by using a close fit as the seal between a graphite piston and a glass cylinder and air as the fluid. The small leak acts as a parallel orifice.

I once invented a zero-stiction dashpot for satellites as part of the Vertistat† gravity gradient orientation device (Fig. 3.7). Sealed tube 1 is filled with silicone oil and contains iron ball 2 as the piston. The ball *rolls* freely inside the tube, and the clearance between ball and tube is the orifice. The ball is coupled to its mechanism member by magnet 3. The dashpot was tested in the laboratory by positioning the tube almost horizontally and letting a very small component of gravity drive the ball. It worked fine both in the laboratory and in the actual satellites. The Vertistat used zero-stiction flexure bearings 4.

There is a class of variable or controllable dashpots in which the orifice is electro-mechanically varied; in effect, the orifice is a servovalve.

---

*Airpot is a trademark of Airpot Corporation.

†Vertistat is a trademark of General Dynamics Corp.

50 Constraint Design

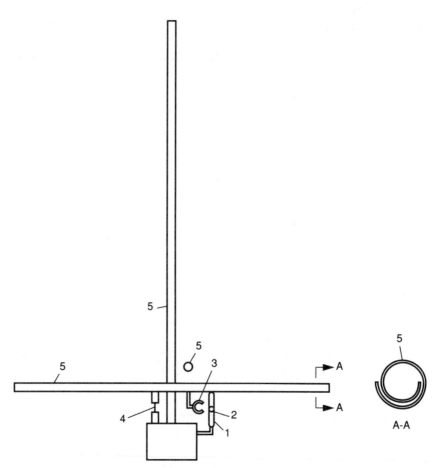

Figure 3.7 Gravity gradient satellite stabilizer.

There is dashpot still in development which uses electro-rheological fluids. The user varies the viscosity of the fluid by varying an electrostatic field through which the fluid flows. In principle, the same effect can be used in clutches and brakes. The development hinges on the development of better fluids. (The magnetic-particle clutch was invented as an electro-magnetic analog to this electro-static clutch.)

Some dashpots are made nonlinear in order to develop approximately constant force at all positions of the stroke as the speed slows down. Artillery recoil cylinders have tapered plugs which gradually close the area of the orifice during the stroke. Some shock absorbers have the equivalent.

3.4.5.2 Cylinders. A pneumatic or hydraulic cylinder exerts a controlled force on the constrained body rather than establishing a defined position; so it may be considered a soft constraint. If the force clamps the body against a hard constraint, the cylinder force becomes a seating force. If the cylinder moves the body between two positions established by hard constraints within or outside the cylinder, then the cylinder is a controllable hard constraint.

A hydraulic cylinder whose control valve is in a feedback loop, either automatic or through a human, provides a variable constraint whose only softness is the compressibility of the oil and the variation in flow through the valve due to variation of load force. For most applications compressibility is negligible.

3.4.5.3 Motors. Motors of all sorts (electric, hydraulic, pneumatic) are controllable soft torque constraints. Servomotors with closed-loop automatic or human feedback constrain to an electrically or hydraulically defined position; they are almost hard constraints except for the softness with which they hold a position. Others apply a continuous torque to their rotating load regardless of position. Still others (torque motors) are used to produce a static torque.

Feedback control systems, whether linear or rotary, are controllable soft constraints. The softness varies inversely with the gain of the servo. The use of rate feedback provides nondissipative damping without the use of dissipative devices such as dashpots.

3.4.5.4 Voice coils. Voice coils are the linear equivalent of electrical dc torque motors.

3.4.5.5 Clutches and brakes. Brakes and clutches are controllable soft torque constraints. They may use dry friction, lubricated friction, eddy currents, magnetic hysteresis, or controllable viscosity (in magnetic-fluid, electro-rheological–fluid, variable-gap, or variable-area devices.) Brakes and clutches are energy-dissipating devices.

"Retarders" are liquid-filled, continuously slipping brakes used to dissipate the energy of heavy trucks as they descend long hills.

Continually slipping eddy-current clutches are controllable torque constraints placed between constant-speed motors and varying-speed loads. The motor and slipping clutch constitute a variable-speed drive. Continually slipping eddy-current brakes are used as controllable drag brakes. Control is achieved by varying the current in the electromagnet, which induces the eddy currents. An analogous device uses magnetic hysteresis instead of eddy currents.

3.4.5.6 Shock mounts. Shock mounts are flexible members with high hysteresis. Each flexible member joins a rigid hub and rigid flange or any other pair of rigid attachment members. The flexible member is usually a block of elastomer or a coil of wire rope. The part to be protected is fastened to the hub, and the flange is fastened to ground. Three or more shock mounts are used to support one body.

A shock mount acts as a combination of spring and damper. It is analogous to an electrical RC filter. Most shock mounts are made of elastomer or knitted wire mesh or coils of wire rope. Some recently developed engine mounts include fluid-filled dashpots. A servo-controlled air spring (a bladder filled with compressed air) is an excellent shock mount. Analysis of vibration modes is very important in both sizing and placement of shock mounts; both textbooks and manufacturers' literature contain analysis instructions.

MinCD is rather unimportant in designing shock mounts or other soft-constraint arrays because their soft spring constants allow easy absorption of dimensional tolerances. Redundant shock mounts may absorb more vibration modes than a MinCD array.

Rubber bushings and grommets through which mounting fasteners pass are equivalent to small shock mounts or mounting feet.

A mechanical filter comprising a stack of alternate springs and masses comprises a very effective shock mount. Such a filter is used as a support for sensitive scientific instruments. A massive version of such a filter is a heavy concrete foundation embedded in earth, but *not* connected directly to the floor of its factory, and supporting a large machine. A more effective version is a heavy concrete block supported by air springs, sometimes with servo control of the air volume in the spring to maintain an accurate vertical position.

3.4.6 Inflated constraint devices

Compressed air is injected into a variety of flexible enclosures to generate soft constraints. The pneumatic automobile tire is the most common.

3.4.6.1 Inflated seal. A seal used in fighter plane cockpit canopies is shown in Fig. 3.8. The hollow tube (a) is inflated with compressed air (b) to make it expand against the canopy edge and produce a seal. (Sealmaster)*

3.4.6.2 Air springs. An air spring is compressed air in a rubber bag. Air springs are used as suspension springs in trucks carrying shock-sensitive loads and also as industrial springs. (Firestone)

---

*Underline means a typical manufacturer.

Kinds of Constraint   53

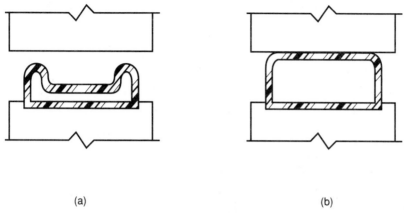

(a)   (b)

Figure 3.8 Pressurized seal.

3.4.7 Part grippers

Part grippers are constraint devices by their nature. Chapter 14 is devoted to robot grippers. Some examples of robot grippers are given here to illustrate the use of soft constraints in grippers.

3.4.7.1 Inflated tube. A robot gripper using compressed air in plastic tubing to grip a row of slippery and fragile ceramic bodies used in catalytic converters is shown in Fig. 3.9.

3.4.7.2 Expanding O rings. A robot gripper in which a pair of O rings acts as an expanding radial piston and stretches outward to grip the inside of a cylindrical part is shown in Fig. 3.10a. Two O rings are used which roll on each other and on the groove sides with almost no friction (b); a single O ring sliding on the groove sides would have high friction which would reduce the working force and would resist returning to its inward position.

3.4.7.3 Vacuum grippers. Vacuum grippers are common in robot practice and are preferred when possible because they are simple and

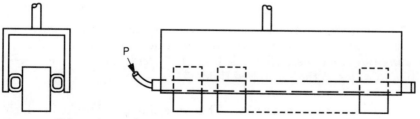

Figure 3.9 Robot gripper tubes.

54   Constraint Design

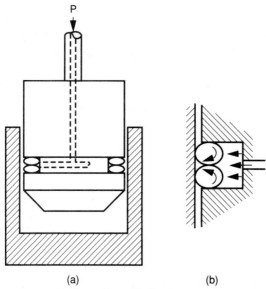

(a)   (b)
Figure 3.10   Robot gripper O rings.

have no moving parts. (One can, with tongue in cheek, think of vacuum as a fluid.) Figure 3.11 shows a vacuum gripper for sacks fed through a large-diameter bellows. The vacuum not only grips the part but can be adjusted to balance the part's weight or to lift it by adjusting the degree of vacuum. (Vacu-Hoist)

3.4.7.4   Ventricles.   In artificial hearts, intermittently pressurized air or water on the outside of a flexible bag (a "ventricle") is a variable constraint which pumps the blood inside the bag. (I built a surgical heart-lung machine using sterile saline solution as the liquid so that a small leak would not foul the blood; current practice for implanted artificial hearts uses compressed air because the tube can be smaller. Having human lives in my hands, like the surgeons, and helping to save them with engineering formed the most frightening and exhilarating experience of my career.)

3.4.7.5   Squeeze valve.   A related device is an industrial valve in which the controlled fluid flows *through* a flexible tube and the *controlling* fluid compresses the tube from the *outside*. The valve is insensitive to solids in the controlled fluid.

3.4.7.6   Air tube clutch and brake.   There is a clutch and brake in which air pressure in a flattened tube exerts force on the rubbing friction

Kinds of Constraint 55

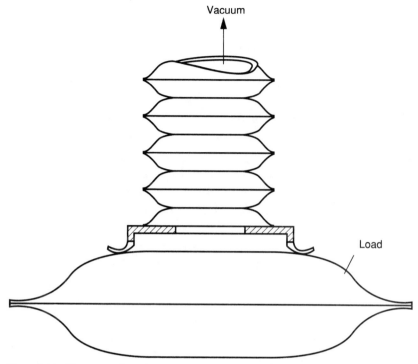

Figure 3.11 Vacuum gripper.

disks or drums as the tube tries to fill out. The tube is the equivalent of a piston and cylinder with no moving parts or sliding seals.

3.4.7.7 Inflatable actuator. There are air actuators made as rubber bags which are alternately filled and emptied of compressed air.

3.5 Flexible Constraints

There is a class of flexible constraints and flexible machine parts other than flexible seating forcers such as springs.

3.5.1.1 Flexures. Flexure hinges, or flexures, are short beams of flexible material which approximate axle hinges through small angles of rotation. They have the advantages of zero looseness, zero lubrication, and zero stiction. They have the disadvantages of not having a precisely defined center of rotation and of having a very limited rotation angle. They are typically made of thin spring metal such as beryllium

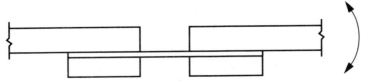

Figure 3.12 Flexure hinge.

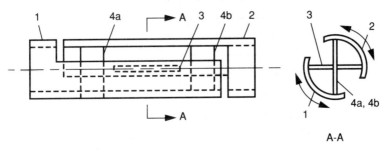

Figure 3.13 Crossed-spring flexure pivot.

copper or spring-temper stainless steel. Figure 3.12 shows a relay armature mounted on a flexure hinge. Many clock pendulums hang from flexure hinges. In Fig. 3.7 the horizontal damping arms of the satellite stabilizer are each carried by a flexure hinge 4.

Figure 3.13 shows a commercial crossed-leaf flexure hinge. Rigid members 1, 2 are clamped in the two parts to be hinged together. Flexure springs 3 and 4a, 4b join members 1, 2. Springs 4a, 4b combined have the width of spring 3 but are kept divided to constrain the freedom to almost pure rotation.

Metal bellows are used as flexible seals in devices to transmit motion into high-vacuum chambers; usually they serve as flexure suspensions for the moving parts.

Many flexure hinges are made of plastic, often molded integrally with the two parts they join together. Polypropylene has extremely long flex life as a hinge and is extruded into long strips for this use.

A standard elastomeric shock mount or bushing can be used as a flexure pivot in both pitch and roll (joystick motion); a pair of such shock mounts constitutes a flexure pivot in pitch only (Fig. 3.14).

Epoxy-fiberglass laminate is used in flexure hinges to support heavy vibrating equipment such as vibratory feeders and conveyors. The proportions used are thick sections and long unsupported lengths so that an S-shaped deformation takes place, equivalent to the motion of a pair of hinges joined by a link (Fig. 3.15).

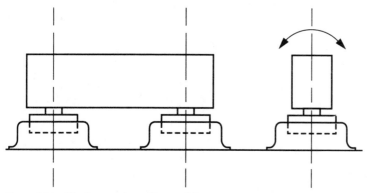

Figure 3.14 Shock mounts as flexure hinge.

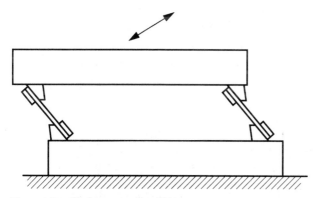

Figure 3.15 Shaker mounting flexure.

3.5.1.2 Flexible couplings. Flexible couplings of many kinds provide angular constraint between a motor and its load. Most comprise a hub on the motor, a hub on the load, and a floating member keyed to both hubs which transmits torque despite parallel and angular misalignment between motor shaft and load shaft. Most have some backlash, which does not matter in unidirectional drives but which may not be permissible in a reversing or positioning mechanism.

Figure 3.16 shows a zero-backlash flexible coupling which uses flexures to transmit torque without backlash despite angular and parallel misalignment. Flexure disks 1 are fixed to hubs 2 and coupling block 3. The flexures act as universal joints with flexure pivots. (Thomas)

Another kind of flexible coupling uses only a single block of elastomer with a metal hub molded or splined onto each end (Fig. 3.17). A crude version is a length of hose clamped to the two shafts with hose clamps. For very light loads a piece of flexible tubing pressed onto the shafts works well.

58 Constraint Design

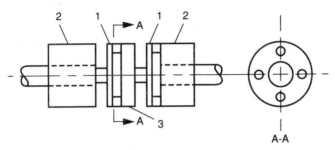

Figure 3.16 Flexible coupling.

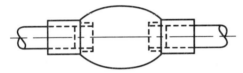

Figure 3.17 Elastomer flexible coupling.

A length of flexible shafting (made of layers of wire helices) is a flexible coupling between hubs which may be close together or many feet apart.

3.5.1.3 Torsion flexures. These have frequently been used to suspend scientific instruments. The flexure is a length of metal wire or ribbon or quartz thread. (Quartz has extremely low hysteresis.) It has zero friction and can be made with a very low spring constant by making it long and thin. Many clocks have been made with torsion pendulums, and some are still made that way for show.

I once used a stiff torsion flexure as a precision angle generator. The problem was to generate angles of a few seconds of arc to test a new kind of gyroscope transducer. We machined an accurate neck 1.000 in long and 1.000 in in diameter in an aluminum bar. I calculated the torsional spring constant of the neck, which was easy to do without approximation. We applied torque inputs and read transducer outputs. A torque accurate to only 10 percent gave angles accurate to 10 percent—*of a few seconds of arc.*

3.5.1.4 Suspension flexures. Rope and chain are used as a suspension flexure. The traditional Japanese bell striker is illustrated in Fig. 2.11. The wire rope suspending the "skull cracker" used to knock down walls is a less artistic example. Cranes and derricks rely on the flexibility of their ropes to permit small horizontal adjustments of the hook position and the load orientation.

3.5.1.5 Bimetal.
A thermostat's bimetal strip is a flexure with its own differential expansion providing the flexing torque.

3.5.1.6 Tape.
Cross-curved metal-tape flexures are used for measuring tapes. They are rolled up for storage and are usefully straight and rigid when released.

For very long satellite antennas and stabilizing booms 5 (Fig. 3.7), beryllium copper is cross-curved through a full turn and a half and then flattened and rolled up for storage. Provision must be made to absorb the stored energy when deploying the rolled-up tape, or it will unroll explosively. Short lengths of commercial measuring tape are used for self-erecting antennas on satellites.

Another form of tape flexure is used as a part-feeder magazine. The original is the machine-gun cartridge belt. A miniature version uses sprocketed movie film with attached pockets to feed semiconductor chips. Intermediate-size sprocketed feeder tapes are increasingly used as part feeders in assembly machines. (I have been trying to promote, for assembly machines, a kind of tape feeder which can accommodate a very large variety of parts.) Tape feeders are much cheaper, more versatile, and much more jam-free than vibratory feeders.

3.5.1.7 Electrical flexures.
One of the great benefits of electric power is the ability to carry it along flexible conductors.

Although it is not used as a constraint, a U-shaped stack of thin copper sheets is a flexure used to conduct large electric currents to a moving slide, such as on a spot or seam welder (Fig. 3.18).

Stranded copper wires and cables of separately insulated wires are electrical flexures used to carry current to moving parts. Division of

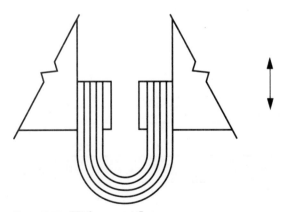

Figure 3.18 High-current flexure.

60 Constraint Design

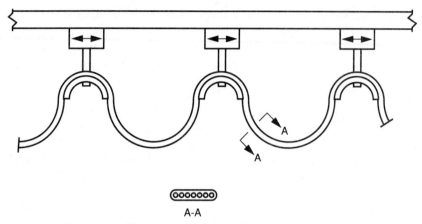

Figure 3.19 Festooned cable.

conductors into strands and laying of strands and conductors into ropelike helices provides flexibility, as with rope.

Ribbon cables, with conductors side by side instead of helically laid, are also flexures which trade off reduced lateral flexibility for greater transverse flexibility. Ribbon cable is sometimes made as a printed circuit strip.

Festooning (Fig. 3.19) is a commercial technique using rolling carriers for carrying ribbon cable and flexible hoses to machine parts, such as crane carriages, moving long horizontal distances. An alternative is reels which pay out and take up long, flexible cables and hoses. (Gleason Reel)

Figure 3.20 illustrates a parallel-motion flexure I designed for an

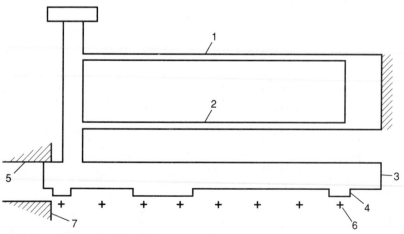

Figure 3.20 Flexure keyboard key.

experimental computer keyboard. It is made of a single piece of metal having parallel flexure portions 1, 2. When the button is pressed, code portion 3 moves downward parallel to its original position and coded tabs 4 interrupt light rays 6. The flexures have a preload against backstop 5 and have a hard forward stop 7 to give a conventional feel to the button.

Many keyboards use diaphragm contacts; the one-piece surface is flexible at each contact area, and pressure flexes the diaphragm inward to make an electrical contact. The system is liquidtight, which makes it ideal for cash registers in fast food restaurants, and all contacts in the keyboard are made as portions of a laminated assembly of insulation, metal, and surface flexure; so cost is very low.

Most electrical connector and relay contacts incorporate the contact and spring into a one-piece flexure. (Larger relay and contactor contacts are rigid and use separate springs.)

Flexible conduit is a standard electrical wiring material.

3.5.1.8 Tension flexures. Stranded rope, both wire and organic fiber, is a mechanical flexure which transmits tension constraint, with five other freedoms, despite bending around sheaves. Helical laying of the strands in the manufacture of the rope permits easy flexure. The derrick ropes of Fig. 2.10 and the suspension chains of Fig. 2.11 are tension flexures. Most gymnasiums have exercise machines with tension flexures to transmit gravity load to straining muscles.

3.5.1.9 Hoses. Hoses and flexible tubes are flexures used to carry fluids to moving machine portions.

3.5.1.10 Flexible hose and cable supports. In some machine tools, wire and hose flexures are tied to a steel or epoxy-fiberglass flexible strip to support them as they flex. This practice has largely been supplanted by threading the flexures through hollow jointed chain links. The link joints constrain the chain to move in one plane only. Some chain is made of metal, and some is made of plastic. A competitive enclosed hollow flexure is also made. In addition to wire and hose, fiber-optic and coaxial cables are carried. (Gleason Reel)

For vertical motion of cables and hoses a simple hanging U with one end fixed and the other rising and falling is usually quite sufficient. However, I have tied a steel wire rope into such a U bundle to support the gravity load, and it once prevented a wreck when an accidental tension overload occurred.

It is essential to prevent kinks in a wire or hose flexure, or repeated bending will break the flexure. A kink is a short-radius flexing portion at an anchor point.

62   Constraint Design

Many mechanical designers tend to relegate wire and hose flexures and stationary wiring and plumbing to the category of afterthoughts or else delegate their design responsibility to assembly technicians. ("Wire and plumb to suit.") In fact these portions of a machine can be among the most challenging to design. Unless designed well, these flexures are the parts most vulnerable to failure and are the subject of expensive last-minute changes and needlessly high manufacturing and maintenance costs. Cabling and plumbing are parts of your machine and need engineering just as any other parts do.

3.5.1.11 Energy storage flexures. Springs are flexures used to store driving energy for some mechanical devices. Most are spiral flat springs like the old alarm clock springs. Model airplanes are still powered by rubber bands in torsion. Ancient catapults used rope in torsion. Ancient and modern bows store the energy to propel arrows with wood, animal-horn, and steel springs. Energy storage is discussed in Chap. 15.

3.5.1.12 Balancing springs. Cross-curved springs are made which deliver approximately constant force throughout a long linear stroke or multiturn rotation to balance the weight of tools and machine parts.

3.5.1.13 Flexible containers. Many containers have flexible portions to enable you to expel their contents. Among these are oilcans, plastic containers, and bellows. (Bellows are used as pumps more often than as storage containers.)

3.5.1.14 Musical instruments. Bells, chimes, tensioned strings, drumheads, woodwind reeds, trumpeters' lips, and tuning forks are all flexures. Except for the reeds and lips, each combines a stiff flexure with a mass to oscillate at a natural frequency and usually at its harmonics also. The reeds and lips couple to the natural frequencies of air columns.

3.5.1.15 Clock crystals. The quartz crystals which have replaced pendulums and balance wheels in timing devices and which establish the frequencies of radio systems flex like musical chimes and have electronic feedback to keep them going. There is a further discussion of mechanical oscillators in Chap. 19.

3.5.1.16 Latches. A common design feature is a cantilever spring latch, which may be integral with the latched part. The latch comprises a cantilever spring flexure with a hook formed in the end. The latch may be designed to open either by forcing it open or by disengaging it with a tab, also formed in the end of the latch.

3.6 Adjustable Constraints

3.6.1 Reasons for adjustability

A constrained position is established either by manufacturing the parts with sufficient accuracy or by adjusting them after assembly. The stability of a position may be sufficient as manufactured, or it may drift in service and require readjustment in service.

I know no straightforward design rules for deciding where to provide adjustability except the universal but not very helpful principle, "It depends." Typically there are tradeoffs among adjustability, stability, manufacturing accuracy, and available maintenance skill.

3.6.2 Adjustable parameters

The parameters which you can make adjustable in your designs include:

3.6.2.1 Straightness. For example, a track which is fastened at many places along its length can be bent into straightness and refastened.

3.6.2.2 Squareness, plumbness, and levelness. Mechanisms and structures can be designed to take advantage of the ease and accuracy with which these parameters can be measured and adjusted.

3.6.2.3 Angle. The angle between parts, other than 90°, can be measured with a protractor, usually built into a tool.

3.6.2.4 Phase. The phase angle between portions of a cycle can be measured and adjusted. Typically this requires adjusting the angular position between two hubs joining two shafts, such as in Fig. 3.21.

3.6.2.5 Distance. Adjustable distances vary from the clearance between sliding parts, which may be less than a thousandth of an inch, to many feet.

3.6.2.6 Alignment and parallelism. You can adjust the alignment of rotational axes and the parallelism of linear paths. "Concentricity" is alignment of rotational axes.

3.6.2.7 Force and pressure. These are adjusted by a linear displacement of a spring or linkage, by adding or subtracting weight, or by

64 Constraint Design

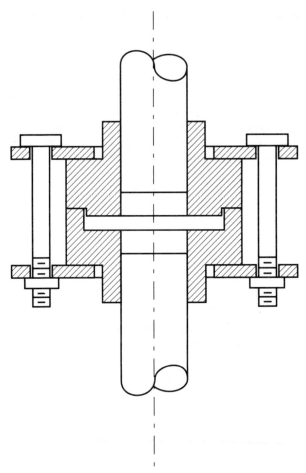

Figure 3.21 Phasing coupling.

adjusting a fluid pressure. Fluid pressure is adjusted by a position displacement on a spring or lever in the pressure controller.

Force can also be generated and adjusted electrically. A modern analytic balance balances the unknown weight with electro-magnetic force. It then measures the electric current which generates that force.

3.6.2.8 Electrical parameters. In electro-mechanical machines, mechanisms are controlled by electrical devices which are themselves subject to adjustment. Typical electrical adjustments are of resistance (potentiometers) and inductance. Increasingly, controls are by electronic computer (microprocessor), and adjustment is effected by changing data entries in its memory.

3.6.3 Adjustment techniques

3.6.3.1 Adjusting screws. Adjusting screws are used as adjustable constraints in leveling surveying instruments and in innumerable other mechanisms. Typically some means is provided to lock the screw in the desired position. Locknuts and lock screws with plastic inserts deformed by the mating screw threads are most common; thread-locking adhesives are highly reliable but impede readjustment; split nuts tightened with cross screws are sometimes used; high-friction nuts (stop nuts) are used; and, in four-screw leveling devices for surveying instruments, a fourth screw serves as an adjustable seating force for the other three.

Figure 3.22 shows two flanges seated against a ball and clamped against it with three screws. When the screws are loose, there are roll and pitch freedoms between the flanges. Differential tightening of the screws adjusts roll and pitch. When the screws are tight, there is no relative angular freedom.

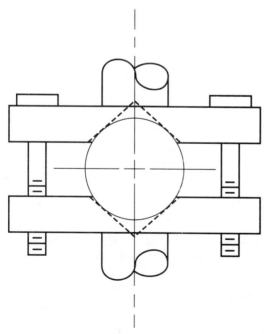

Figure 3.22 Angle adjustment.

It is important to know the difference between a position-adjusting screw and a seating setscrew. The "setscrews" on a dovetail gib are really position-adjusting screws, not setscrews. The screws which fasten a pulley hub to a shaft are setscrews. (See Sec. 3.8, "Friction Constraints.")

3.6.3.2 Eccentrics. Eccentrics are used as hard-constraint adjustments. A useful example is the eccentric on the shaft extension of some cam followers (Fig. 2.3). An example of use is Fig. 2.15.

3.6.3.3 Shims. Shims are used to adjust hard constraints. The work of adjusting is tedious, but shims can take large forces and do not loosen or drift.

3.6.3.4 Wedges. Wedges and wedge pairs are used to adjust hard constraints and to serve as sources of seating force. A screw and nut are a helical wedge pair.

3.6.3.5 Phasing hubs. The hub of Fig. 3.21 permits adjustment of the rotation angle between the shafts. Note the central relief, which assures that the pressure between the hubs is at a substantial radius so that the torque capacity is high.

3.6.3.6 Adhesives and grout. Adhesives and grout are used to lock an adjustment after it has been made by other means; those other means may then be removed.

3.6.3.7 Hammers and files. These are precision tools in the hands of a skilled craftsperson and just what you would expect in the hands of a roughneck.

3.6.3.8 Part replacement. Certain parts may be made in increments of size and the best size chosen for each assembly ("selective assembly"). Parts may also be made in matched sets and each assembly made from such a set. Balls in high-precision ball bearings are made and chosen in this way.

3.6.4 Adjustment measurements

Adjustment is a feedback control process: you vary the adjustment until a feedback instrument tells you that you have achieved the results you wish. The instruments used in adjusting mechanisms include the following.

3.6.5 Geometrical instruments

3.6.5.1 Level. Bubble levels (Brown & Sharpe) and electronic levels (Celesco) vary from rough carpenters' tools through precision instruments with a resolution measured in seconds of arc. Some levels are applied directly to the work, and some are used to establish a level optical axis for a telescope or laser which has optical targets applied to the work.

3.6.5.2 Square. Mechanical squares are made in the same range of accuracies as levels. Pentaprism optical squares are used to bend a light beam at exactly 90°.

3.6.5.3 Angle scales. Protractors and telescope rotation scales measure angles. Machine-shop dividing heads use worms and worm gears. Sine bars measure angles with great accuracy and fine resolution. Linear measurements combined with trigonometry measure angles. (In ancient Egypt they set up right angles for land surveying with 3-4-5 triangles made of rope.)

3.6.5.4 Plumb line. A string or piano wire and weight (plumb bob) indicates the vertical with great accuracy. To measure deviation from the vertical of a surface (Fig. 3.23) I made twin plumb lines 1, used a

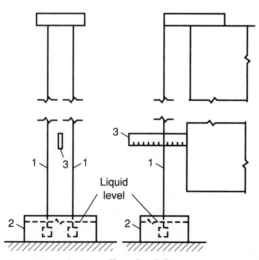

Figure 3.23 Antiparallax plumb lines.

pot of water 2 to damp swinging of the weights, and used a scale 3. The setup let us measure deviation of a surface from the vertical to $\frac{1}{32}$ in with the naked eye. The twin plumb lines enabled us to read the scale without parallax. We were easily able to adjust a vertical rail 20 ft high, in two directions, to this tolerance. You can replace the scale with a telescope on a fixture for much tighter tolerances.

3.6.5.5 Linear-distance instruments. Depending on the distance and the desired accuracy, there is a large range of instruments for measuring linear distance. These include:

- Steel tape and scale
- Caliper
- Micrometer
- Gauge blocks
- Feeler gauges
- Dial indicators (mechanical, pneumatic, electrical)
- Machine tool optical scale (Sony)
- Precision rack and pinion
- Ultrasonic range finder (Polaroid)
- Optical "radar" (Cubic)
- Laser interferometer (Hewlett-Packard)

3.6.5.6 Optical measuring systems. Originally developed for land surveying and then refined for the aircraft industry, systems of very accurate optical measuring instruments are available. They are based on light rays either *to* optical telescopes or *from* lasers and use targets, scales, protractors, right-angle prisms, and the optical analog of radar. The laser interferometer (listed above) is an optical distance-measuring instrument using laser light and electronic counting of light-wave interference fringes.

Polarized light may be used to transmit an angle over long distances. A polarizer is attached to one part, and an analyzer is attached to the other. When the light through both is extinguished, the parts are parallel in one freedom.

3.6.6 Other instruments

3.6.6.1 Human touch, hearing, and vision. It is surprising how much can be learned from careful looking, listening, and touching without instruments. Included are smoothness of motion, looseness, vibration, and alignment. In debugging a mechanism, much can be learned by comparing the sensations in your two hands and in your eyes and ears. Even your nose can sometimes be very useful in detecting the effects of heat.

3.6.6.2 Transducers. Inside and outside a machine are physical quantities related to its operation and adjustment: temperature, pressure, speed, radiation, noise, electrical quantities, etc. Transducers of many kinds exist to measure these quantities, and you should become familiar with those which are relevant to your work.

At this point I would like to suggest that you include features in your designs which enable instruments to be coupled to your mechanisms for initial debugging, for manufacturing adjustment, and for maintenance diagnosis.

3.7 Variable Constraints

A *variable* constraint differs from an *adjustable* constraint in that the variable constraint changes during the operation of the mechanism while the adjustable constraint is changed only during manufacture or maintenance of the mechanism. Examples are:

3.7.1.1 Cams. A cam is a variable hard constraint. In automatic screw machines and in four-slide presses a stack of cams on a common shaft or on gear-coupled shafts constrains and synchronizes several outputs from a single input.

Cams are made as variable-radius rotating disks, variable-length rotating barrels, and variable-height linear slides. There are textbooks on cam design.

At one time three-dimensional cams were used in control mechanisms for fire control and other functions to deliver a function of two independent variables, $z = f(x, y)$, but they have been largely replaced by tables of numbers in electronic digital controls. A 3D cam was a somewhat cylindrical part which was rotated by one input and translated along the axis of rotation by a second input. The output was a cam follower with a ball tip, which was constrained along a radial axis. The radius of each point on the cam positioned the follower to a point which was a function of rotation and of translation.

When a variable constraint is required to follow a fixed cycle, the old-fashioned cam and follower are accurate, fast, reliable, cheap, strong, electronics-free, and easily understood by your customer.

3.7.1.2 Linkages. When the geometry permits, a mechanical linkage may be even better than a cam and follower. Four-bar and other linkages are a subject in machine kinematics and are extensively treated in the literature. Section 2.7 describes the whiffletree linkage used to distribute a constraint to many points on a body.

3.7.1.3 Lead screws. A lead screw and nut are a matched set providing a variable hard constraint. Lead screws are used in machine tools and in instruments. They are turned by servomotors following electronic programs, by gears from a machine tool spindle drive as in a screw-cutting lathe or a power-feed milling machine, or by hand. Lead screws have either acme threads and solid nuts or ball-bearing threads and ball-bearing nuts. MinCD lead screws in which the nut has a single point of contact with the screw are made for instruments. (Flennor)

Usually a lead screw and nut are designed with RedCD coupling to their driven slide. Great care and accuracy are used to make the axis of the screw parallel to the axis of the slide and to mount the nut so that its axis coincides with the axis of the screw (Fig. 2.5). Any imperfection results in friction and in bending of the screw.

Figure 3.24 shows a MinCD connection between the nut and the slide which eliminates screw bending and unnecessary friction. Nut 1 is coupled to gimbal 2 by tie rods and rod ends 3. Gimbal 2 is pivoted in extension 4 of slide 5. The gimbal and rod end bearings prevent parasitic torques between the slide and the nut (i.e., provide angular freedoms). Long torque arm 7 in slide slot 6 prevents nut rotation without significant bending force on the screw. Backlash is prevented by preload nut 8 and spring 9. (See Chap. 16 for a discussion of backlash and its elimination.)

This problem of coupling a lead-screw nut to a slide is a common one. I am by no means confident that my solution is the best because of the number of parts it uses. I hope some readers will take this problem as a challenge in MinCD and let me know their solutions.

3.7.1.4 Air cylinders. An air cylinder provides a variable hard stop. Because air is compressible, air cylinders are generally used to establish only two positions established by hard stops within or outside the cylinder. Multiple-stop positions for air cylinders can be achieved by inserting one or another of a set of hard stops ("interposers") into the path of a piston-rod attachment (Fig. 3.25). I have made multiposition robot axes in this way.

3.7.1.5 Hydraulic cylinders. The elastic modulus of oil is very high, so that for most purposes oil may be considered incompressible. Thus hy-

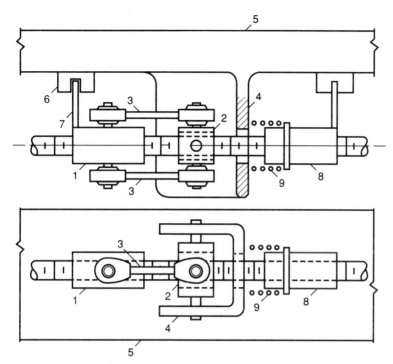

Figure 3.24 Lead screw to slide MinCD.

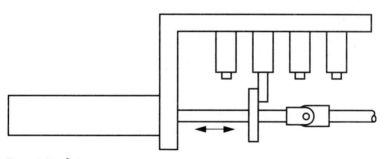

Figure 3.25 Interposers.

draulic cylinders with appropriate control valves provide a hard-stop position anywhere along the cylinder stroke. (Since oil is actually somewhat compressible, the stop position established by a cylinder moves slightly under heavy load. For hydraulic servos requiring high-frequency response, the compressibility of the oil may cause harmful resonant frequencies.)

3.7.1.6 Electric motors. Electric motors provide the following classes of variable constraint:

1. *DC and induction motors* apply continual torque during rotation but do not establish a defined position.

2. *Synchronous and stepping motors* establish a defined angular position as a function of time.

3. *Servomotors* in feedback control systems establish a defined angular position as a function of another parameter which may vary with time.

4. *Linear electric motors* position a slide directly without intervening mechanism such as screws or gears.

3.7.1.7 Gears. A gear train is a variable hard constraint on the angle between two shafts. If there are differential gears in the train (bevel or planetary), the constraints relate three or even more shafts. A worm and worm gear are used to vary the angle between a device and its supporting structure; examples are the azimuth and elevation gears of telescopes, antennas, and cannons.

3.7.1.8 Rack and pinion. A rack and pinion are a variable hard constraint equivalent to a lead screw. The pinion can be stationary and drive a moving rack, or the rack can be stationary and the pinion, carried by the load, move along it. The latter combination is used in preference to long lead screws on most mechanisms which travel many feet because of the sag, Euler column strength limit, cost, and elastic compressibility of long lead screws.

Racks of unlimited length may be assembled from short lengths by properly spacing the adjacent ends. We routinely assembled racks over 100 ft long. A short length of rack can be used as an assembly tool by meshing its teeth with the teeth of adjacent lengths until those lengths are fastened down. As a design tip, I suggest that you orient racks with teeth down if possible or sideways as a second choice. If the teeth point up, they will collect dirt on the pressure surfaces.

A pair of parallel racks with mating gears on a common shaft (Fig. 2.15) is a variable constraint which operates on one linear freedom and one angular freedom; i.e., it generates a parallel motion.

Figure 3.26 shows a special case of an X-Y table driven by stationary gears. Table 1 carries rack pair 2, 2a for the X-axis drive and rack 3 for the Y-axis drive. The X racks mesh with elongated gear 4 (actually commercial pinion rod), and the Y rack meshes with elongated gear 5. Each rack is driven along one axis by its gear and slides freely along its gear on the other axis. The gears are driven by servomotors 6, 7.

Backlash elimination and other aspects of rack and pinion drive are described in Chap. 16.

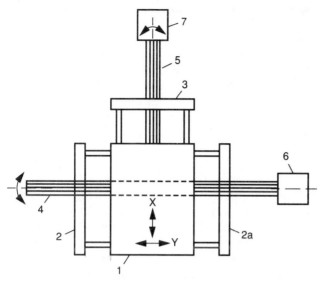

Figure 3.26 X-Y table.

3.7.1.9 Chain, tape, and rope. Chain, tape, or rope driven by a sprocket or drum and an electric motor is a variable hard constraint having low cost, long travel, and high strength. A second chain or the same one in a closed loop provides a seating force. But beware of the secret trap: a long length of chain or steel tape or wire rope is elastic. At some positions its spring constant, combined with the mass of the moved body, may have a natural frequency within the control range of the control system and result in intolerable resonances. Furthermore, elastic stretch affects the calibration of any system which measures position by measuring the rotation of the sprocket or drum.

3.7.1.10 Belt and pulley. Belt and pulley constitute a class of variable angular constraint between two rotating shafts. Included in this class are smooth belts and pulleys, toothed belts and pulleys, chains and sprockets of many kinds, and V belts and pulleys. Smooth belts and pulleys use friction and can slip or creep so that an exact angular relationship is not maintained. Other belts use a toothed connection between "belt" and "pulley" analogous to the chain and sprocket so that an exact angular relationship is maintained.

Smooth belts stay on crowned pulleys in the same way that tapered railroad wheels stay on tracks without constant dependency on flanges and their friction and shock. The only difference is that the pulley is stationary axially and the "track," which is the belt, moves axially. The two sloping sides of the pulley's crown work as the two cones of the wheels do.

3.7.1.11 Servos. Many variable constraints are driven by electric or hydraulic motors or cylinders in a feedback control system. The input to the control system may be a template, or stored computer data ("numerical control"), or a pilot track such as an embedded wire in the floor for a factory automatic guided vehicle, or an inertial guidance system, or an automatic pilot, or other control input.

3.8 Friction Constraints

Static friction, under the seating force of gravity, enables us to walk without skidding and prevents tables and chairs and machines and cars from sliding around in response to small forces. Even when a body is bolted down, the bolt tension serves only to increase the seating force, and it is friction which exerts the constraint.

A rolling wheel has the same friction constraint as a stationary wheel until a skidding force breaks it loose from the friction. A soft wheel, such as a pneumatic tire, obeys this same rule, but since it deforms under lateral force as it rolls, its path is complicated. Most single wheels exert little yaw constraint because the contact area is small and so the moment arm from friction point to friction point is small. Two wheels fixed to a common axle a substantial distance apart exert large yaw constraint.

The friction at each point of contact between the ground and a body exerts two linear axes of horizontal constraint, east-west and north-south, and the contact exerts the third, linear, vertical axis of constraint.

Bodies held together with fasteners are constrained by friction, with the fastener tension providing the seating force. An exception is a doweled connection in which the dowel constrains by direct interference. A second exception is a hot-riveted connection; the rivet expands to fill the two holes while hot and in cooling provides tension so that it also constrains by direct interference.

It is important to be aware of the difference between a friction constraint, which can slip, and a positive-interference constraint, which cannot.

3.8.1 Friction devices

Clamps of many sorts provide seating force to generate high friction between parts. Examples are:

3.8.1.1 Collets. Collets, split and tapered, are wedged between a shaft and a hub by one or more axial screws to produce both centering and friction between the shaft and the hub. Collets are used to couple hubs to shafts and to hold workpieces in lathes. Solid elastomeric bushings are used in this manner to clamp and seal electrical cables.

3.8.1.2 Chucks and vises. These tools hold parts by friction.

3.8.1.3 Wedges. Wedges and wedge pairs are used to adjust spacing and to clamp parts. The wedges are usually retained by friction, although some wedges are moved and retained by adjusting screws.

3.8.1.4 Taper pins. Taper pins are wedging dowels.

3.8.1.5 Setscrews. Setscrews are used as seating forcers. They generate friction constraint between a hub and a shaft and between a key and its keyway. Many setscrews have cone points or serrated hollow cone points to cut into the shaft; these serve as positive-interference constraints and also resist unscrewing.

3.8.1.6 Belt and pulley. Belt-and-pulley drives rely on friction between belt and pulley. Since the belt can creep and slip, their constraint is not geometrically positive, as are chain and gear drives. (Tooth or timing belts engage sprockets and resemble chain.)

3.8.1.7 Friction variable-speed drives. A class of mechanical variable-speed drives uses friction between rollers, balls, and other shapes rolling on variable-radius surfaces. Another class uses friction between a V belt and variable-radius pulleys.

3.8.2 Screw thread retention

Screw threads are prevented from unscrewing by friction, and an elaborate art is devoted to establishing and maintaining that friction. Some of the techniques used are:

3.8.2.1 Plastic inserts. Threads are formed in the plastic by the engaging threads. The plastic threads, so formed, grip the engaging threads elastically.

3.8.2.2 Self-tapping screws. These hardened screws cut or form their own mating threads, which therefore fit tightly.

3.8.2.3 Deformed threads. Deformed threads are pushed back into nominal shape by the entering mating thread and therefore grip the entering thread by their own elasticity.

3.8.2.4 Tapered threads. These threads are used in pipe. They wedge tightly as they are torqued up.

3.8.2.5 Adhesives. Adhesives are introduced between mating threads. These special-purpose adhesives are available in a broad range of strengths. They drag during the entire unthreading operation. They are also quite valuable as an assembly means between hubs and shafts. For example, in Fig. 2.15 we joined the pinion stems to the long coupling shafts with an adhesive made for thread locking. There was zero backlash and no stress concentrations, wall thickness was small, and cost was low. To disassemble, one merely decomposes the adhesive with heat.

3.8.2.6 Lockwashers. Single-tooth and multiple-tooth lockwashers provide extra friction between a screw head or nut and the adjacent joined body. (*Design note*: It is a common error to put a flat washer between the lockwasher and the joined body to prevent the lockwasher from scratching the body. The effect is to bond the flat washer to the screw head or nut and *not* to prevent the flat washer, together with the screw or nut bonded to it, from rotating freely.)

3.8.2.7 Locknuts. Locknuts provide extra friction between a nut and a screw. They are also used to clamp adjusting screws.

3.9 Self-Aligning Elements

There are many ways in which pivots can provide a desired combination of linear constraints and rotary freedoms. These freedoms permit mechanisms to align parts and axes under the working torques which appear, and they permit parts to be self-aligning. This statement is pretty abstract; so let's look at examples.

3.9.1.1 Caster. The common wheel caster (Fig. 3.27) is a self-aligning device. The vertical-axis pivot allows the drag force to torque the

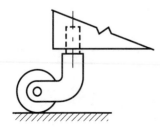

Figure 3.27 Wheel caster.

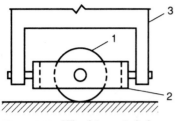 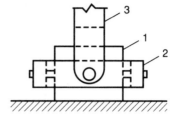

Figure 3.28 Wheel in a gimbal.

wheel around until it rolls in the direction of motion. (It is not universally known that if the direction of motion is reversed, the caster will swivel 180° without any lateral motion of the vertical axis being required. The required lateral motion is provided by the rolling of the castered wheel.)

3.9.1.2 Single gimbal. A wheel in a single gimbal is shown in Fig. 3.28. The gimbal enables the wide wheel to maintain a full line of contact with the ground despite variations in the angle between the supported member and the ground.

3.9.1.3 Two gimbals. A two-axis gimbal (having two rotary freedoms) is shown in Fig. 3.29. Such a gimbal set is used to provide freedom of roll and pitch between a mariner's compass and the motion of a ship. Compass 1 is pivoted on the roll axis in gimbal 2. Gimbal 2 is pivoted on the pitch axis in gimbal 3. Gimbal 3 is fixed to ground. In a compass mount there is yaw constraint; so the compass body always has the same yaw orientation as the ship. The bearings have axial constraint (not shown); so there are a full three linear constraints.

3.9.1.4 Two gimbals (alternate). A different two-axis gimbal set is shown in Fig. 3.30. Instead of using cantilever stub shafts, it carries its load in direct cross compression along the full length of each shaft. Its load capacity is quite large, but its rotation is limited. It is used to carry the thrust of steerable rocket engines.

3.9.1.5 Three gimbals. A three-rotary-axis (i.e., three rotary freedoms) gimbal set is shown in Fig. 3.31. Shaft 1 is free to roll, pitch, and yaw, but it has no linear freedom with respect to ground 2. The "fifth wheel" trailer hitch in Subsec. 2.5.1.12 is an example, although the construction of its axes looks different.

78 Constraint Design

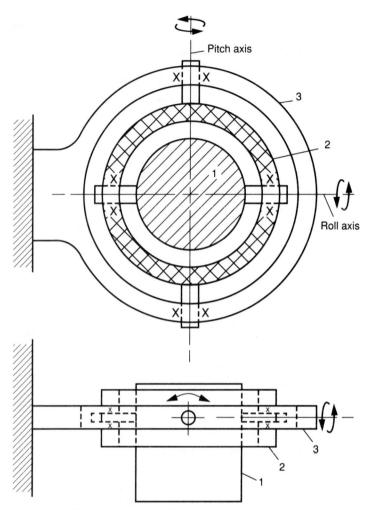

Figure 3.29 Two-rotation gimbal set.

The three-gimbal assembly of Fig. 3.31 transmits linear force without any rotary constraint. Compare it with the assembly of Fig. 2.2, which transmits rotary torque without any linear constraint.

3.9.1.6 Sliding gimbals. A two-rotary-freedom and two-linear-freedom gimbal set is shown in Fig. 3.32. It has only two constraints: rotation and axial translation along the central axis. The two linear freedoms are the gimbal trunnions sliding endwise in their bearings. I have seen this combination used in a very elegant manufacturing machine.

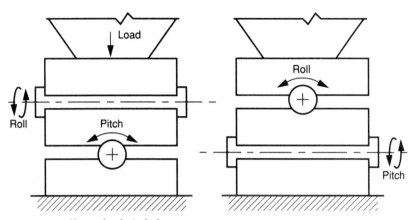

Figure 3.30   Heavy-load gimbal set.

3.9.1.7 Universal joints. The conventional universal joint ("U joint," "Hooke's joint," "Cardan joint") is shown in Fig. 3.33. It is equivalent to the compass gimbal of Fig. 3.29 except that the central cross 3 carries both sets of pivots. The arrangement is intended to provide constraints between the shafts 1, 2 in relative roll, X, Y, and Z with freedoms in pitch and yaw. Section 3.5 describes U joints with flexure pivots. The Rzeppa joint is a more complicated universal joint which maintains uniform angular velocity across a single joint and is therefore used in front-wheel drives in automobiles.

3.9.1.8 Self-aligning linear ball bearing. The self-aligning linear ball bearing made by Thomson Industries is illustrated, in principle, in Fig. 3.34. It was used by my robot company in the linear-motion modules of Figs. 2.12 and 2.13. Each row of balls has a pivoted outer race so that it can self-align with the shaft pitch and yaw. In the fully closed version (Fig. 3.34) the bearing inside diameter accurately matches the shaft outside diameter; so they comprise a nonadjustable matched set. In the split version (Fig. 3.35) there are adjusting screws on the bearing diameter; so the bearing and the shaft comprise an adjustable matched set.

3.9.1.9 Spherical joint ("ball joint"). A sphere in a spherical socket is used in many self-aligning devices as a matched set to provide three rotary freedoms with three linear constraints. The spherical rod end (Fig. 3.36) is a very widely used example. Ball 1 is partially enclosed in closely fitting socket 2, each being a partial sphere. The ball has a hole and two flats so that it can be bolted to one of the parts to be joined. The socket 2 has extension 3, which is threaded into (or onto)

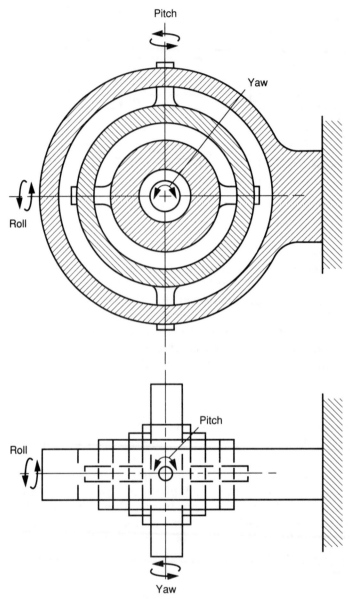

Figure 3.31 Three-rotation gimbal set.

the other part to be joined. Linear forces are transmitted between the joined parts, but all three angular axes are freedoms. Thus no torques are transmitted because of structural misalignments or mechanism motions (Tuthill). The ball-joint trailer hitch in Sec. 2.5 is similar in principle.

Kinds of Constraint 81

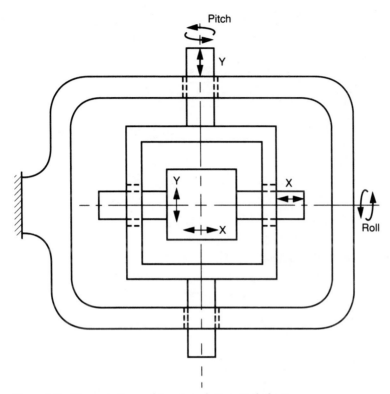

Figure 3.32 Two-rotation and two-translation gimbal set.

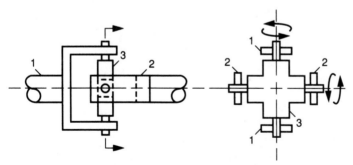

Figure 3.33 Universal joint.

3.9.1.10 Spherical bearing mount. Spherical pillow blocks (Fig. 3.37) are rolling bearings (ball or roller) the outside of whose outer race is a partial sphere fitted into a spherical socket in the pillow block. The axis of the bearing is free to self-align with the axis of the shaft. This requires only two of the three angular degrees of freedom of the spher-

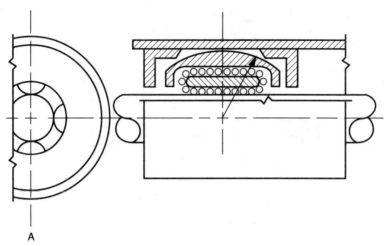

A

Figure 3.34 Self-aligning linear ball bearing.

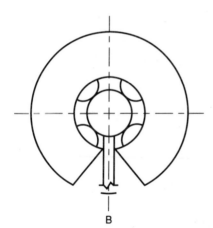

Figure 3.35 Linear ball bearing with gap.

B

ical joint. The third degree of freedom either is ignored as harmless or is positively keyed to prevent rotation and wear.

3.9.1.11 Leveling pads. Spherical joint leveling pads (Fig. 3.38) have an attachment stud or tapped hole on the ball and a flat surface (foot or pad) on the socket. They permit a body to rest on a support surface on areas instead of points, which is a common semi-MinCD practice. Leveling pads are standard tooling components.

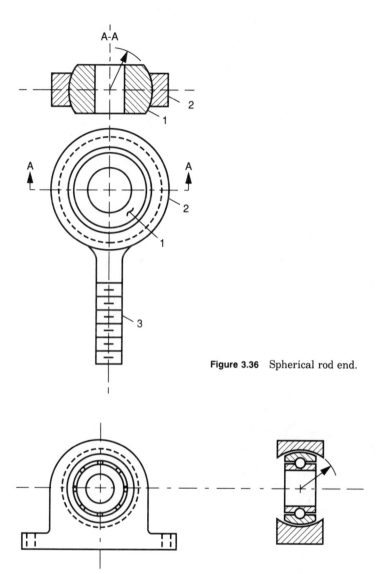

Figure 3.36 Spherical rod end.

Figure 3.37 Spherical pillow block.

3.9.1.12 Spherical bearing. A spherical bearing (Fig. 3.39) is the same as a rod end without any extension of the socket. It provides a sleeve bearing, either between the shaft and the hole or between the ball and the socket, with the same self-aligning freedom as the spherical pillow block.

84 Constraint Design

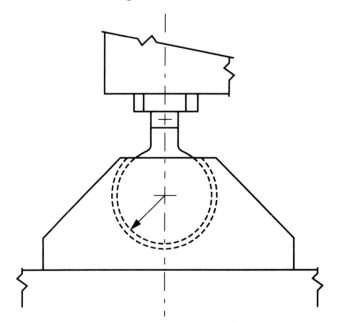

Figure 3.38 Spherical leveling pad.

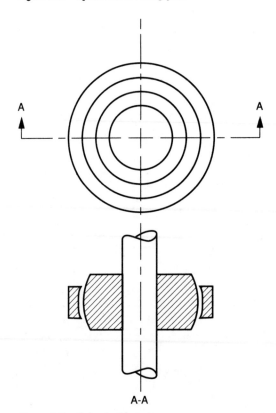

Figure 3.39 Spherical bearing.

Kinds of Constraint   85

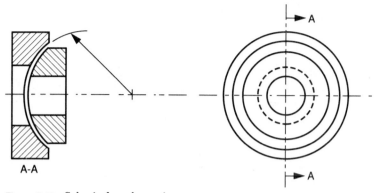

Figure 3.40 Spherical-washer pair.

3.9.1.13 Spherical washers. Spherical-washer pairs (Fig. 3.40) are pairs of washers which have matching ball-and-socket surfaces. If placed between two nonparallel surfaces, they permit a bolt to pull the surfaces together with area contact on each. You can make what I call a "poor man's" spherical-washer pair in which the seat is a cone and the "ball" is the end of a cylinder with a rounded chamfer (Fig. 3.41). Geometrically these should touch at only two points, but if they are tightly forced together, the rings deform to ellipses with a full line of contact and the contact line deforms to a contact band. I have made large-diameter pairs so that a large-diameter lead screw could pass through, with the washer pairs used to make self-aligning seats for the lead-screw bearings.

The pure solution for the problem of Fig. 2.4 in which rigid feet must be bolted to a mounting surface is to provide spherical-washer

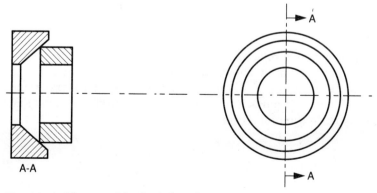

Figure 3.41 "Poor man's" spherical washers.

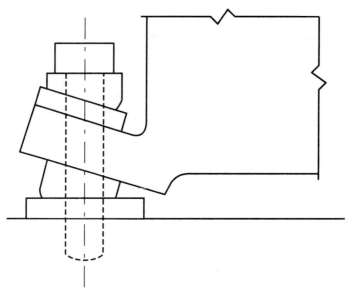

Figure 3.42 Foot with spherical washers.

pairs between each foot and the mounting surface *and* between each foot and the bolt head (Fig. 3.42).

Spherical-washer pairs are standard components in tooling catalogs.

3.9.1.14 Self-aligning roller bearing. A form of self-aligning roller bearing (Fig. 3.43) has the outer race spherical, the rollers barrel-shaped, and the inner race shaped like a barrel with a concave side. The rollers roll with a slight skid, as do the balls in a ball bearing. The roller path on the outer race self-aligns with the axis of rotation, and the inner race oscillates if the axle is not perfectly on the axis of rotation. These bearings are used for service under very high vibration in which the shaft flexes from the vibration forces.

3.9.1.15 Linear chained roller bearing. The chained rollers of the Roundway* roller system of Thomson Industries Inc. is shown in Fig. 3.44. The chain of rollers self-aligns between the cylindrical track and the cylindrical core. All parts are ground to accurate dimensions and therefore constitute a matched set.

---

*Roundway is a trademark of Thomson Industries Inc.

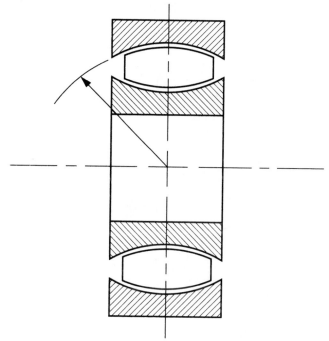

Figure 3.43 Self-aligning roller bearing.

3.9.1.16 Ball caster. A ball caster (Fig. 3.45) uses a ball instead of a wheel. The ball rolls on a quantity of small loose balls in the spherical portion of the housing and can change its direction of rolling without the intermediate displacement required by a wheel caster.

3.9.1.17 Spline. A spline (Fig. 3.46) has a linear freedom and an angular constraint; so it transmits torque between parts whose distance varies. Splines are also made with recirculating balls between the parts to reduce the friction in the linear motion.

After reading all this you deserve a story.

Company A manufactures artillery shell cases. There is a threaded hole in the flange for a fuse, and the thread has a close-tolerance specification.

When I visited, as a consultant, the threads were inspected with go and no-go thread gauges. There was no problem with the go gauge. If it was tight, the company forced it through and cold-formed the thread

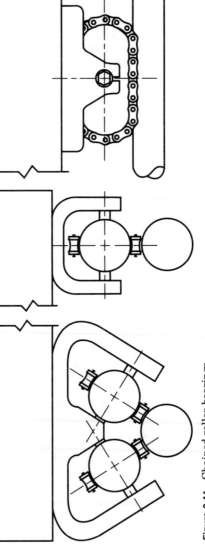

Figure 3.44 Chained roller bearings.

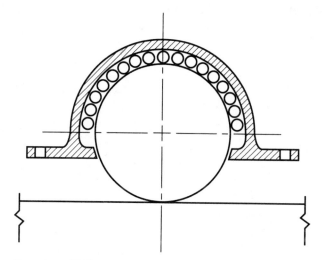

Figure 3.45 Ball caster.

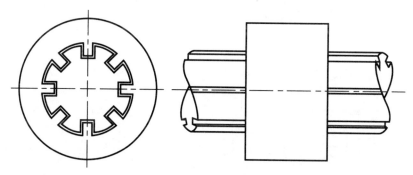

Figure 3.46 Spline.

to take it. This was not the government's intent, but the result was satisfactory as long as worn gauges were replaced often enough.

There were plenty of problems with the no-go gauge. The company inspectors never got them through, but on a bad day the government inspectors did. After a no-go gauge went through once, the part was permanently lost since the gauge opened up the thread beyond the tolerance even if it should not have gone through in the first place.

I asked what the gauge torque specification was and how the company measured it. "A light twist of the wrist," the company said and would not budge; so I looked it up. *NBS Handbook H28*, the screw thread bible, actually said just that. I could hardly believe it.*

---

*The specification is now FED-STD-H28. In the issue dated March 5, 1985, Sec. 6A, Par. 4.1.1, it says, "Gaging decisions are based on the torque applied by the inspector when assembling the gage on the product thread...."

90 Constraint Design

I built the company an automatic no-go thread gauge in accordance with simplified drawing 3.47. The thread to be gauged 1 is in the flange 2 of the cartridge case. No-go gauge 3 has extended stem 4, whose diameter is the root diameter of the thread so that the gauge self-aligns with the thread to be gauged. Gauge 3 hangs from universal joint 5, spline 6, and universal joint 7 so that it has freedoms on all axes except rotation about its axis (compare Fig. 2.2).

The assembly is carried by bearing pair 8A and is coupled to large-diameter gear 9.

Coaxial with gear 9 is identical gear 10 on bearings 8B and driven

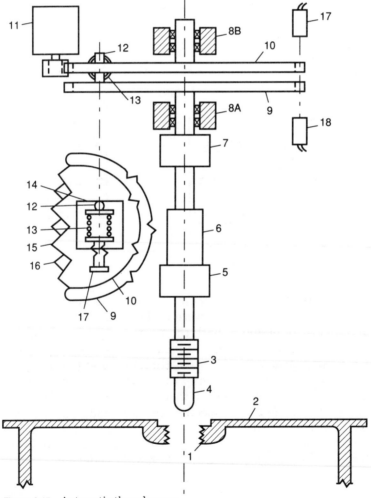

Figure 3.47 Automatic thread gauge.

by motor 11. Gear 9 is coupled to gear 10 by pin 12 on gear 9, which is urged by spring 13 on gear 10 against surface 14 on gear 10. *Spring 13 is adjustable and is set to produce the desired no-go torque; if that torque is exceeded, pin 12 lifts off surface 14 and there is relative rotation of the two gears.* The positions of pin 12 and surface 14 are such that when they touch, the teeth of the two gears are out of phase and prevent the light of lamp 17 from reaching photocell 18.

In action, the cartridge case is lifted against the gauge, and the gauge tries to screw into the thread. If the torque exceeds the spring-set torque, the gears rotate relative to each other and the photocell responds. The thread is declared good, and the motor is reversed. If there is no photocell response before the gauge penetrates a specified amount, as indicated by limit photocells not shown, the thread is declared bad and the motor is reversed to back out the gauge.

The device worked fine, and the customer's manufacturing engineers were so delighted with it that they kept it protected in a closet and used it only as a showpiece while they continued to fight with the inspectors over the "slight twist of the wrist." To actually introduce an innovation was too much for them. The real world of manufacturing engineering! (See Chap. 11.)

3.9.1.18 Active self-alignment. We have been considering devices in which self-alignment occurs as the result of interference forces. In some machines, such as robots, self-alignment on selected axes is achieved by feedback control of their axis drives. (See Chaps. 13 and 14.) An extreme case is a military homing missile which "self-aligns" to the X, Y, Z of its target by controlling its own roll, pitch, and yaw.

Chapter 4

Beneficial Non-MinCD

4.1 Semi-MinCD

Absolutely pure MinCD can be achieved, its benefits are dramatic, and the designer's satisfaction is profound. However, pure MinCD has limitations which have been presented in many parts of this text. Usually some degree of compromise with absolute purity has negligible bad effects and is economically justified. (It is just as foolish to be rigidly doctrinaire about MinCD as about most other policies.) Design with such compromises is semi-MinCD.

4.2 Matched Sets

A class of semi-MinCD is matched sets. They take advantage of standard commercial parts made with very close tolerances. Within the matched set the design is RedCD, but its high manufacturing accuracy requires only very small elastic deformations for the matched set to operate. Among the available matched sets are:

1. Rolling bearing
2. Screw and nut (both sliding and ball)
3. Spline
4. Pair of cam followers on the sides of a rail (Fig. 2.3)
5. Ball-and-socket rod end (Fig. 3.36)
6. Ball-and-socket self-aligning bearing race (Fig. 3.37)
7. Linear-motion ball bearing on a hardened and ground round rail (Fig. 3.34)
8. Linear-motion multiple-race ball-bearing slide and track
9. Linear-motion roller-bearing block between flat ways
10. Sleeve-bearing pair and shaft
11. Meshed gears in an accurately made gearbox

94   Constraint Design

12. Rack and pinion
13. Chain and sprocket
14. Concentric sets
    a. Shaft and hub
    b. Rotor and stator of motors, etc.
    c. Fluid power cylinders
    d. Solenoids and voice-coil actuators
15. V-edge wheels and mating tracks
16. Spherical-washer pairs (Fig. 3.40)
17. Ground sliding ways

4.3 Finite Area Contacts

It is semi-MinCD to use finite area contacts in an otherwise MinCD configuration instead of using point contacts. There will initially be a point of contact within each finite area. Then some combination of elastic deformation, inelastic deformation, and wear deformation will enlarge the point contact to an area contact. Many accurate machines, such as internal-combustion engines, rely on wear deformation ("running in" or "breaking in") as the final manufacturing process.

4.4 MinCD to Semi-MinCD Conversion

4.4.1.1 Conversion of Fig. 2.1. Figure 2.1 shows an example of pure MinCD. Suppose the light spring seating force were replaced by a heavy load. The six point contacts would crush until sufficient area of support developed. The semi-MinCD approach is to shape the abutting surfaces so close to each other that only elastic deformation is needed to support the load. Leg 4 becomes a matching cone, leg 5 becomes a matching wedge, and leg 6 becomes a matching flat. "Matching" is easy to say, but it may be costly to actually machine the parts to be sufficiently "matching" to support their loads elastically.

4.4.1.2 Heavily loaded slide. A useful combination for heavily loaded sliding surfaces, as in machine tool ways, is to provide a self-aligning matched set such as the gimbals of Fig. 3.30 between the load and a sliding matched set of ways.

4.4.1.3 Lathe carriage. Many lathe carriages rest on three areas, two on a front way and one on a back way (Fig. 4.1), each extending many inches along the ways. Thus there is a combination of a stable tripod support and the load-bearing capacity of large-area ways. I have also seen a large lathe headstock bolted to the bed casting with three large feet, a semi-MinCD tripod mount.

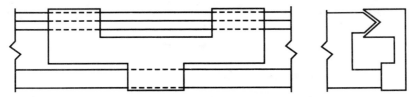

Figure 4.1 Semi-MinCD lathe carriage.

4.4.1.4 Bolted foot. Although I have given a theoretical example of the use of two pairs of spherical washers as the pure MinCD way to bolt down a foot without stress, it is usually less expensive to design the foot region strong enough to withstand mounting stress, accurate enough so that the amount of stress generated by the bolt is within the strength of the part, and with a big enough bolt that it can withstand its off-center load. Then three such feet constitute a semi-MinCD tripod mount.

4.4.1.5 Zero-looseness hinge. A zero-looseness hinge is illustrated in Fig. 4.2. It is used to pivot a displacement transducer in an instrument which measures semiconductor wafers to 20 μin. Rotating arm 1 turns on shaft 2 in stationary block 3. The shaft engages both the arm and the block in V grooves 9. The shaft is tightly clamped to the block by clamps 4. It is spring-seated against the arm by springs 6 and screws 7. End play is eliminated by seating spring 11 and washer 8. When the arm rotates, it slides around the shaft despite the friction produced by seating springs 6. Mechanical stops 10a, 10b establish the working position. This is a semi-MinCD device; the two regions under clamps 5 approximate two points of a three-point constraint, the stops 10a, 10b are a pure third point, and the axial spring 11 and washer 8 approximate a fourth constraint point. The shaft bends elastically to accommodate the minute imperfections in the V grooves.

4.5 Useful RedCD

Although a purpose of this book is to introduce MinCD and encourage its use, its broader purpose is to make you a better designer with insight into the relative merits of all three approaches: MinCD, semi-MinCD, and RedCD. This section is devoted to examples where RedCD is clearly superior to MinCD.

4.5.1.1 Large distributed load. When a part is subject to a large and distributed load, for example, a cylinder head, it is economical to provide a redundant number of distributed fasteners instead of sizing the part to carry the load to MinCD fasteners.

96　Constraint Design

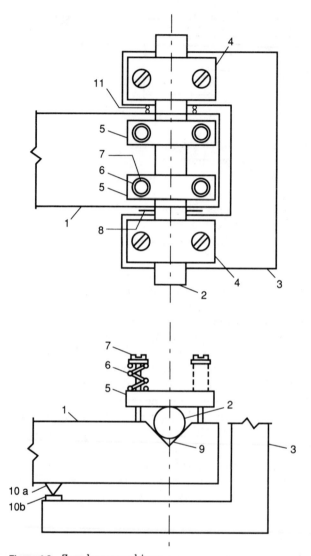

Figure 4.2　Zero-looseness hinge.

4.5.1.2 Necessary deformation. If a part must be deformed to be satisfactorily assembled to its mating parts, a RedCD set of fasteners is the appropriate constraint system. Fluidtight flanged joints are an example.

4.5.1.3 Varying load distribution. When the loading of a system varies, it may be useful to provide redundant constraints to accommodate changing loads. An example is a machine with wheeled carriage running on a track having joint gaps. Redundant wheels cross the joint gaps smoothly since when one wheel is over the joint gap, another wheel carries the load and prevents the first wheel from dropping into the joint gap. This is illustrated in Fig. 16.6.

4.6 RedCD Components

There are a number of valuable mechanism components which are inherently RedCD. Examples are:

4.6.1.1 V-band fastener. This is sometimes referred to as the Marman clamp because it is made by the Marman Company. It is illustrated in Fig. 4.3. Most of these fasteners are made of sheet metal rolled to

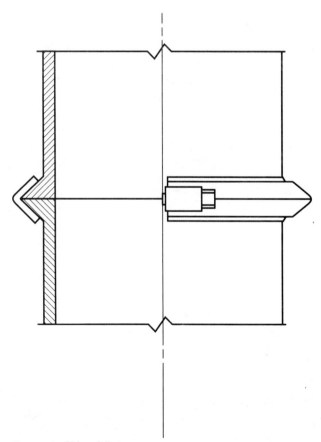

Figure 4.3 V-band fastener.

shape and are inherently inexpensive. There is no diameter limit. The two ends of a length of clamp are joined by a bolt, which tightens the band around the parts to be joined. The band elastically conforms to the small variations of shape and diameter of the parts it joins so that it easily overcomes the tolerance disadvantage of RedCD.

In my earlier book, *Successful Engineering*, as an example of simplification in design I proposed a V-band clamp for the space shuttle solid-rocket-booster joints. The shuttle now uses a monstrously expensive and complicated set of 180 radial dowels in blind square-bottomed holes. An experienced rocket engineer took me to task and said: "V-band clamps had been investigated for the purpose and absolutely cannot be made to work, period. The cases are made of hardened steel and are stressed to their safety limit to minimize weight. If a correspondingly thick V band were made, the stress raisers at the internal corners would cause it to fail. It can't be done, see, and don't bother me anymore."

Back to the drawing board. (See Fig. 4.4.) The band is coupled to the shells through the sides of two round rods inserted tangentially; so there are no sharp internal corners as stress raisers. The rods are inserted tangentially through holes in the band. All parts are machined; so although there is no wedging together as with a V band, the looseness can be made negligibly small and is tightened by internal pressure as soon as the rocket is ignited. The band is divided by one or more cuts tied by bolts to permit assembly; there is no spreading force because of the wrap around the intermediate rods. The O-ring configuration is handbook. (The *Challenger* disaster was caused by narrow O-ring grooves which prevented internal pressure from pushing the O rings into the gas escape path, which is how O rings work. See Ref. 21.) I invite controversy.

4.6.1.2 Retaining rings. The retaining ring is a simple, compact device to resist axial force. Both the ring and its groove are inexpensive. Figure 4.5*a* shows an internal ring, and Fig. 4.5*b* shows an external ring. Both are split, elastically contracted or expanded to pass the main diameter and enter the groove, and held in the groove by their own elasticity. They are made in an enormous range of sizes and materials and have no inherent size limit. The manufacturers make simple installation tools.

Retaining rings are subject to the RedCD limitation that if the mating parts are not accurate, load concentrations may make the assembly fail. However, since both the groove and the face of the abutting part are lathe-cut, they are almost inevitably a very close match.

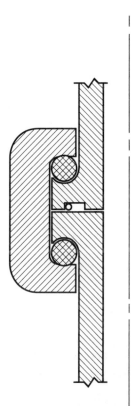

Figure 4.4 Heavy-load band clamp.

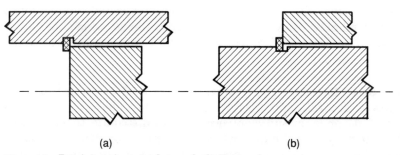

Figure 4.5 Retaining rings. (*a*) Internal. (*b*) External.

100   Constraint Design

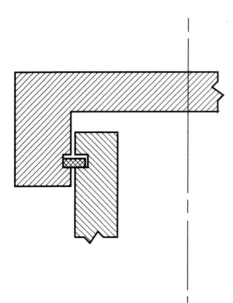

Figure 4.6 End-inserted retaining ring.

Some retaining rings are made with a beveled edge which, when it engages a matching bevel on the side of its groove, acts as a spring-loaded axial wedge and clamps the retained part against a shoulder at its other end.

Figure 4.6 shows a retaining ring which joins two cylindrical parts. The ring is inserted endwise through a window in the outer part and is pushed endwise around and through the matching grooves. Some hydraulic cylinder heads are retained in this way, making tie rods unnecessary.

Figure 4.7 shows how two retaining rings can provide an axial grip with zero end play. They replace a machined flange, which not only is expensive but may interfere with assembly. I have used this arrangement successfully.

4.6.1.3 Screw threads. The screw thread is the ultimate in RedCD. It depends on accurate manufacturing even to be able to be assembled and then depends on sufficient accuracy to distribute the load over several turns; yet the screw thread is the most common mechanism element I know. Different standard tolerances provide different uniformities of load distribution.

The RedCD limitation shows itself in the economical limits within which screw threads are made; I know of no threads as large as 1 ft in diameter except for the breechblocks of very large cannon.

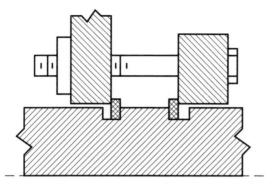

Figure 4.7 Retaining rings providing zero end play.

4.6.1.4 "Piano hinge." The continuous hinge is an elegant special case of RedCD in which the multiple hinged joints not only constrain the two parts to a single relative freedom but constrain their edges to be parallel along the entire length of the hinge. A long flexure hinge does the same. (S & S Hinge)

4.6.1.5 Flanged joint. The flanged joint is a standard configuration. Even if the flanges depart from both round and plane, a RedCD bolt circle will bring them together. There is no diameter limit.

4.7 Self-Improving RedCD

There is a class of RedCD between parts which slide on each other in which wear deformation continually *improves* the accuracy of the parts.

4.7.1.1 Wearing in. I have already referred to the internal-combustion engine in which the initial operating period wears off roughness of the parts.

4.7.1.2 Flat lapping. The flat-lapping process in manufacturing wears off high spots on both the workpieces and the lap, both getting progressively better.

4.7.1.3 Parabolic lapping. In the manufacture of parabolic telescope mirrors, the shape is generated and continually improved by wear between the mirror and the tool.

4.7.1.4 Bearing-ball lapping. Finish lapping of ball-bearing balls is done in the tool of Fig. 4.8. Since the V races are circular, the radius of

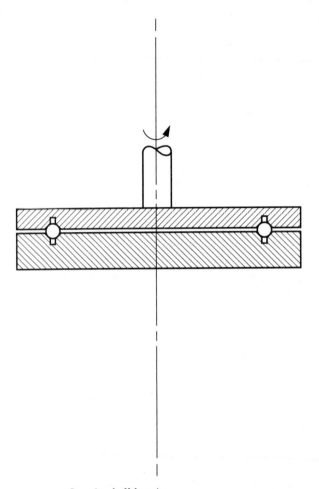

Figure 4.8 Lapping ball bearings.

the outer ring of contacts is larger than the radius of the inner ring. Therefore, the balls must slide as well as roll, and lapping occurs. The larger balls carry most of the load and wear faster. The high spots on all balls carry most of the load and wear faster. The high spots on the tool races also wear faster, of course; so both tool and product become more and more uniform as the process continues.

4.7.1.5 Circle divider. The crown-gear circle divider is a commercial product which gets more accurate the more it is used (Fig. 4.9). There is no central bearing; the two parts are constrained together only by the fit of their tapered teeth under gravity as the seating force. The tool is not continuous; it divides the circle into only as many parts as

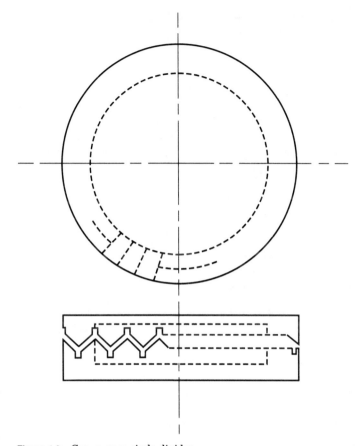

Figure 4.9 Crown-gear circle divider.

there are teeth. The more it is used, the more nonuniformities are worn off.

4.7.1.6 Lead-screw lapping. Lead screws and nuts wear themselves into uniform contact but do not correct the lead variation of the screw along its length. To make an extremely uniform lead screw, a split nut is made which is as long as the screw and is spring-seated inward against the screw all along its length. The screw is run back and forth a short distance, and all nonuniformities are worn off, including lead nonuniformities.

4.7.1.7 Hand scraping. In the manufacture of machine tool ways, the surfaces are sometimes hand-scraped. Bluing the way and then rubbing it with a flat tool reveals the high spots to the worker, who then scrapes them off. The tool itself improves with time.

4.7.1.8 Conical bearings. Conical journal bearings self-improve by wearing off high spots. The bearings move axially to take up wear.

Exercises in MinCD, Semi-MinCD, and RedCD

1 Sketch three designs of each of the following devices. First make a design based on conventional RedCD, then one based on pure MinCD, and then one based on semi-MinCD.

    Knee-type milling machine
    Lathe
    Microscope
    Reclining chair
    Bed equivalent to a water bed (*Hint*: Use whiffletrees. Omit RedCD.)
    Other devices of your own choice

2 Design MinCD and semi-MinCD lead-screw assemblies different from those shown in this book. Include the lead-screw bearings and their supports.

Part 2

Designing with Commercial Components

Chapter 5

General Discussion

5.1 Commercial versus Special

Should you use commercial components in your machine, or should you design your own?

5.1.1 Advantages of commercial components

5.1.1.1 Development costs. There is no cost for engineering, testing, or tooling.

5.1.1.2 Manufacturing costs. Manufacturing costs may be less than for specials because the quantities manufactured include those for many other customers as well as for you.

5.1.1.3 Experience. Commercial components have already been field-tested and refined; so reliability may be higher than for a new design.

5.1.1.4 Approvals. Commercial components may already be on a list of products tested and approved by a recognized laboratory (particularly electrical products), or approved by your customer, or approved by your own company.

5.1.2 Advantages of your own design

5.1.2.1 Suitability. You may suit your own needs more closely than a commercial product can. (But remember that component manufacturers are usually willing to make you a modified version or even a complete special.) The requirements for a machine tool, a fighter plane, and a consumer automobile are rather different.

5.1.2.2 Costs. If your quantity needs are large, you may be able to save cost overall.

5.1.2.3 Design integration. You may be able to integrate the component into other portions of your machine instead of coupling it in. You may thus save space, weight, part count, and cost.

5.1.2.4 Independence. You get a feeling of independence from vendors. If you are human, you may get an ego trip. But you are exposed to criticism for reinventing the wheel; so have your defense ready. Search your soul for your *real reasons*.

5.1.2.5 Management considerations. Your management may have considerations other than optimum design, such as keeping another engineering or manufacturing department busy.

5.1.2.6 Combining ideas. You may copy and combine ideas from several competing vendors together with your own. (Just be sure you don't infringe any patents. If unpublished ideas are disclosed to you in the hope of selling to you, decide what your own ethics are in using them. Be warned that you may be committing an "unfair business practice." I once sued a major machine tool company for doing just that with my robot ideas and proposals, and it settled by promising not to use them, which was all the justice I could afford.)

In either case, you will want to know about the commercial components which are available, so let us proceed.

5.2 Approved Products

Unfortunately there is no industrial version of the Consumers Union that you can consult for its test results and recommendations. However, there are some lists of tested and approved products which you can consult (in some cases, which you had better consult).

5.2.1 Your company

Your company stocks certain commercial components in the inventory for its existing products, and you may save it money if you try to design with them.

Your company may also catalog common shapes which it has already made for other machines (hubs, flanges, etc.). You may benefit your project both in engineering and in manufacturing costs by using these existing parts. The buzzwords for this practice are "group technology."

Your company also stocks certain raw materials and semifinished materials such as sheet steel in certain gauges, shafting, etc. You will be able to get models made faster, save the company money, and increase your own popularity by designing around these materials.

Your company may or may not have formalized a written preferred-components list. If it has, you can follow it or undertake to have it changed. For example, fasteners are a category in which there exist thousands of commercial varieties and in which you may usually choose more than one that are equally appropriate for each use.

Your company has a variety of other design standards which you should learn, again to prevent having to reengineer.

"Your company" is actually many people with different points of view. The purchasing department may have different preferences from the engineering department, which may have different preferences from the manufacturing department. Presumably you work for the chief engineer and will take your direction from him or her.

Your company may also have a blacklist of materials, components, and vendors that it does *not* want to use for a variety of reasons. If you inadvertently call for such a material, component, or vendor, you will have to reengineer.

5.2.2 Other organizations

There are two electrical testing organizations which test electrical products and list those which meet their specifications. They will also test your new design, but the cost in money and, especially, in time may be onerous. They are:

Underwriters Laboratories Inc. ETL Testing Laboratories Inc.
333 Pfingsten Road   Industrial Park
Northbrook, Illinois 60062   Cortland, New York 13045
312-272-8800   607-753-6711

5.2.3 Your customer

The military have qualified-product lists which you can get from your contracting officer as part of the specification package governing your project. They most certainly have a large number of specifications governing what you design for yourself.

Large industrial companies have similar lists and specifications. I have found that different plants in the same corporation may have different lists and specifications.

There are many horror stories. I know a major machine tool company that delivered and installed a quantity of large precision lathes which worked fine. It then got to completely rewire them, in the customer's plant, because its own wire color code was different from that in the customer's specs. *Then* it got paid.

5.3 Sources of Information

5.3.1 Your program of study

I'm sorry to use such a threatening expression, but it is essential. A major part of your continuing informal postacademic study is learning components: what there is (old and new), where to get them, how much they cost, and how reliable the vendors are.

5.3.1.1 Catalogs. I recommend that you *build a catalog library* unless your company library is *really* convenient. I suggest that you use library boxes, not file drawers, and that you file by subject, *not* by company name. If a catalog contains different products, cut it up or get duplicate catalogs. You can then, as a design process, pull out the appropriate box and scan all the relevant catalogs to find the solution to your need without tedious searching.

I particularly recommend two sets of catalogs for every mechanism designer. Catalogs in the first set have substantial technical texts and are sources of education in their field. The manufacturers named in this book are sources. They are referred to throughout the book with their names underlined. To avoid unnecessary space consumption, manufacturers' names are condensed to the essential portion of those names which will enable you to find the entire names and addresses in a directory.

The second set consists of what I call "department store" catalogs, each with a wide variety of products. These are listed as Refs. 72 through 84 in Chap. 20.

5.3.1.2 Advertisements. Read the ads in the trade journals. The good ads are much better written than most of the articles and are full of useful information on components. These are not consumer ads; the advertisers use them to sell by teaching you their products. References 58 through 61 are journals I recommend.

5.3.1.3 Trade shows. Go to trade shows. The ads come to life, and you can ask questions. You may find yourself talking to the exhibitor's chief engineer. An enormous amount of serious engineering is done in trade shows in the midst of those people who are just goofing off on a paid holiday.

5.3.2 Purchasing directories

There are many directories to help you find components and their manufacturers. They are listed in the References as items 63 through 66. Your company library should have some, and your public library has several. Many are too expensive for you to buy for yourself. Your local Yellow Pages is a directory and can be a big help in finding local sales offices.

5.3.3 Manufacturers' representatives and salespeople

Reps vary from good to bad. They do have the advantage of being local, having a catalog and maybe a sample when they visit, and being rather quick to respond. Some really are useful application engineers for their principals. If you have a question for the factory, try to bypass an information intermediary. I have been speaking of commission reps. Employed salespeople and sales engineers employed by vendors are usually more useful, and some are enormously helpful as design consultants—free. (I encourage engineers to become sales engineers; I have done a great deal of sales engineering and found that it can be, professionally, a very productive life.)

5.4 Big Companies versus Small Companies

I know of no general differences between big and small companies as either vendors or customers, and I have dealt with many of both. In each category there are those who are honest and dishonest, rigorously high in quality and slipshod, up to date and old-fashioned, high tech and low tech, diligent and lazy, reliable and unreliable. Bigness carries prestige but not necessarily deserved prestige; I have known companies that coasted for years on prestige earned by a generation long since retired. Part of your responsibility is to form a judgment of the company as well as of the product.

References 67 through 71 are business directories which will give you business information about companies. You will find them in the public library and some of them in broker's offices. (You will find it interesting and perhaps valuable to look up your own company in these directories.)

5.5 Components in This Book

There is a breathtaking number of products manufactured. The products discussed in the following chapters are only those which are components of machines. Products which are *not* included are:

Complete machines

Manufacturing devices

Laboratory devices

Maintenance products

Tools

Office machines

Supplies

Packaging products

Electronic components whose specific function is not visible to the mechanism designer (speed controls, in; transistors, out)

Materials (You can't learn enough about materials, but there is too much information to include materials in this book. See Refs. 31, 32, 39, and 61.)

5.6 Organization of This Book Section

Inventors and engineers do not create devices to fit categories in a neat, preestablished array. In my classification I try to organize components into a pattern which helps a mechanism designer to find what he or she needs. It is unlikely that there is any perfect pattern. I have established one I think is helpful, but I am sure there will be arguments. (The word for organizing into classes and subclasses is "taxonomy"; biologists do it a lot. Knowing the word hasn't helped me at all.)

There is an all-encompassing classification which has been established by the federal government for combining statistics on manufactured products and which is widely used for some business purposes; it is the Standard Industrial Classification (SIC) codes (Ref. 42). You should know about it, but I do not think it helps as a design guide.

Please go to the table of contents and scan it from Chap. 6, "Rotary Motion," through Chap. 9, "Other Components." This will give you an overall view of the classification scheme. The following chapters follow this scheme.

5.7 Breadth and Depth

To describe the variety of components and to give technical data on each would require a large encyclopedia. Reference 38 is the *Mechan-*

ical Components Handbook, which gives a great deal of technical data on the components it describes but necessarily limits the variety of components it covers. The approach of this book is to describe the variety of available commercial components as completely as I can so that you will know what you can get, but it leaves the data to be found, as a second step, in the manufacturers' catalogs.

If you know a trademark or a brand name of a product, you can find its manufacturer in the American Trademark Index of *Thomas Register* (Ref. 65).

5.8 How to Use This Book

When you have a general idea of the kind of component you need, read the corresponding section to learn more about what is available. The next steps are to consult a directory to learn the corresponding manufacturers, acquire their catalogs, consult their sales engineers, see and test samples, consult other users for their experiences, and finally design it in.

A way to find the category you need is to start with a company name and look in the cross-reference section for its listing. Other companies making similar products will have adjacent listings in the same category.

The names of companies in this book are given partly because of the value of their catalogs and partly to give you a start in a directory search. This process is complicated by the variety of products made by large companies; you will have to search the listings until you find the category you want.

If you know a trademark or a trade name, the directory may have a section telling you the name of the company that owns it. (*Thomas Register* does.)

I find it useful to write a one-page specification for the product I seek and send it, with a brief cover letter, to a quantity of manufacturers I check on the pages of a directory. I give a draft of the specification and letter and the check-marked directory to a secretary with a word processor, and I am finished with that task until the replies come in.

This process assumes a formal design sequence, which is not always the way life really is. Often, your knowing of the existence of a component steers your design thinking into benefiting from and needing that component. Therefore this book is really a textbook which should be read through *before* you start a design.

This book is not a directory. Manufacturers are named because their products are unusual, or because their catalogs and publications are useful learning aids, or because their names will aid in a directory search. *This use of names is not an endorsement that a named manufacturer or its products are better than others*. That decision must be made by you, and it is often not an easy one.

Chapter 6

Rotary Motion

6.1 Bearings

6.1.1 Rolling bearings

Good teaching catalogs in this field are published by Kaydon, SKF, Torrington, THK, NSK, Fafnir, Timken, and other bearing manufacturers. These catalogs provide detailed design data for bearing selection and sizing.

6.1.1.1 Ball bearings. Ball bearings are made for shaft diameters ranging from less than $1/16$ in to more than 2 ft. They are configured as:

1. Single-row bearing having a small axial looseness.
2. Single-row four-contact bearing having zero looseness.
3. Angular-contact single-row bearing having large single direction and axial as well as radial load capacity.
4. Matched set of two angular-contact bearings having mutual preload and zero axial looseness (Fig. 16.1).
5. Two-row bearing capable of supporting a spindle by itself.
6. Thrust bearing. Some large-diameter thrust bearings are built as turntables. (Kaydon)

Some bearings contain commercial balls and separators and sometimes one commercial race, with the remainder made integral with the customer's parts by the customer. Gyroscope rotors are an example.

Wire race bearings use hard steel wires inserted into grooves in the user-built housing for the balls to roll on.

Large turntables can be built economically by using large commercial balls rolling in user-built races of unhardened steel. (For a given race diameter, larger balls give more load capacity than smaller balls.)

For high temperatures, ball bearings are made with ceramic and sapphire balls and races.

Bearing balls are separated by conformal metal spacers, or by undersize balls, or by phenolic rings having individual ball holes and also serving as oil sponges.

The ball space is either unprotected, relying on the installation to exclude dirt and provide lubrication, or is shielded against dirt by a thin metal dam or sealed to retain lubrication grease by a rubbing elastomeric ring, or both.

The cross section of the races varies from very small, with small-diameter balls, where space is at a premium as in large, slowly turning shafts, to large-cross-section races with large-diameter balls where the shaft is heavily loaded.

Lubrication is by permanent grease pack, periodic regreasing, oil wick, oil pumping, or oil mist. (See Sec. 9.7.)

6.1.1.2 Roller bearings. Roller bearings provide a nominal line contact at the rolling surfaces instead of the nominal point contact of ball bearings; they offer more load capacity in the same envelope. The penalties are that the rollers must be caged against skewing and that special arrangements must be made to carry axial load.

Roller bearings are made in shaft sizes from ⅛ in to several feet.

Rollers are made in diameters from ¹⁄₁₆-in "needles" (Torrington) to several inches. They are made in cylindrical, tapered, and barrel shapes (Fig. 3.43). Cylindrical rollers carry radial load only, tapered rollers carry both axial and radial load but must be provided in pairs if it is necessary to carry axial load in both directions, and barrel-shaped rollers are used in spherical self-aligning races and have low axial load capacity.

Thrust bearings are made with tapered rollers to provide a nonskid geometry. (Timken)

6.1.2 Bearing housings

Bearing races are made of expensive hardenable steel alloys and therefore are made with a minimum of mounting features. Bearing manufacturers offer a variety of housings and mountings to position and support their bearings in users' machines. (Browning)

One class of such housings, illustrated in Fig. 3.37, is a flange or pillow block having a spherical cavity in which an outer race with a matching spherical outer surface is free to self-align with its shaft.

Similar pillow blocks are made without the self-aligning feature. Pillow blocks may include lubrication and position-adjusting features.

Special bearing nuts are made to clamp inner races to shaft shoulders. They have fine pitch threads and locking-tab washers.

Some bearings have eccentrics to clamp their inner races to their shafts. (Without clamping, the shaft tends to roll along the inner race and fret. Thread-locking adhesive is a good preventer of such action.) (Loctite)

6.1.3 Sliding bearings

6.1.3.1 Hydrodynamic lubrication. The principle of hydrodynamic lubricated bearings is studied in basic machine design courses and is not reviewed here.

Commercial bearings are made of metal, plastic, sapphire, and graphite. (Eagle-Picher)

Metal bearings are made with a variety of soft metal surfaces facing the shaft. Some have a steel back for strength. Sintered-powder bearings are made of bronze or aluminum powder (Oilite) and are porous enough to hold oil and to act as an oil wick.

Kingsbury thrust bearings use tilting shoes to generate the hydrodynamic film against a flat disk. They are made in very large sizes to carry the thrust of ship propellers and the weight of hydroelectric turbine generators.

Some lubricated phenolic bearings are used.

Sapphire bearings are made in instrument sizes. For example, mechanical watches use jewel bearings. (Bird)

6.1.3.2 Hydrostatic lubrication. Spindle assemblies using air hydrostatic lubrication are sold for applications requiring extreme rigidity, low friction, and high speed. Most air hydrostatic bearings are built by the user, but Dover makes commercial units.

6.1.3.3 Dry. Unlubricated bearings are made of plastic or graphite. The principle plastics used are nylon, Delrin, and Teflon,* but many others and many combinations are used. Most are made as bushings, and some have flanges to support end thrust.

Thin split plastic bushings are made which assume the diameter of the shaft and bore, consume little space, are unaffected by shrinkage or expansion of the plastic, and are cheap because they contain little material.

---

*Delrin and Teflon are trademarks of E. I. du Pont de Nemours & Co.

Metal and plastic bearings are made with spherical races similar to rod ends; so they are self-aligning.

6.1.4 Flexure bearings

The subject of flexures is discussed in Sec. 3.5.

Bendix makes a commercial flexure bearing as shown in Fig. 3.13.

6.2 Spindle Assemblies

Complete assemblies containing a spindle, bearings, a base member, and sometimes lubrication components are sold as integrated components for rigid, balanced, high-speed, low-runout grinders and the like. (Whitnon, Pope)

6.3 Coupling Hubs to Shafts

Hubs of pulleys, gears, sprockets, handles, cranks, cams, and other machine elements must be fastened to shafts.

6.3.1 Interference couplings

The cheapest coupling is a setscrew, preferably bearing against a flat. Two setscrews at 90° are much less likely to loosen, and they remove the looseness between the shaft diameter and the hub hole diameter. Commercial setscrews are made with hollow-cone points, knurled, to resist loosening, and with cone points to dig into the shaft or to enter a cross hole, to provide keying. In general, setscrews burr shafts and have much less torque capacity than other devices described below. Setscrew coupling has no angular looseness; so it is suitable for reversing service.

Woodruff keys provide positive keying and are trapped by the hub.

Rectangular keys provide positive keying but must be retained either by a blind-end keyway or by a setscrew. Neither Woodruff nor rectangular keys provide axial positioning of the hub, nor do they eliminate looseness due to clearance between the shaft diameter and the bore diameter. They do not prevent angular looseness and therefore are unsuitable for reversing service.

Cross pins key hubs inexpensively and are particularly useful in freezing an adjusted position of a hub. For large torques, cross pins may have insufficient cross section to carry the load.

Commercial cross pins include straight dowels, grooved or deformed straight dowels, taper pins, and expanding-spring pins of several types.

Cross pins of soft metal serve as shear pins which will break at a certain load to limit torque transmission between hub and shaft. They are the mechanical analog of an electrical fuse.

6.3.2 Tapers and collets

Matching tapers on a shaft and a hub, together with an axial nut to hold them engaged, make assembly easy, eliminate looseness and eccentricity, and provide high-friction torque. A positive key can be provided on the taper.

At least two companies, Ringfeder and Fenner Manheim, make taper bushings which fit between a cylindrical shaft and a cylindrical hub bore. When a circle of screws are tightened, some tapered members wedge inward and mating tapered members wedge outward.

Many pulley and gear hubs for power transmission have integral shaft collets in addition to keyways. In at least one design the collet-tightening screws are transferred to other holes to become collet-loosening screws.

6.4 Collars and Retaining Rings

Axial position can be provided by split friction collars which are tightened around a shaft. (Ruland)

Retaining rings are elastically deformed in diameter for installation. They snap into shaft grooves (external rings) or bore grooves (internal rings). (See Fig. 4.5.) Some retaining rings are made with a bowed shape which provides an axial spring force. Some have beveled edges which wedge against matching bevels on the groove to provide axial clamping. (Waldes Truarc)

Retaining rings are made in diameters from ⅛ in to many inches.

Figure 4.7 shows how two retaining rings can be used as a flange without axial looseness in either direction.

In general, collars and rings replace machined shoulders. This practice both saves money in material and in machining and permits designs which would be impossible with integral shoulders.

V bands couple cylindrical parts end to end. (Marman)

6.5 Shafting

Commercial shafting in mild steel is made to a tolerance of plus zero and minus a small amount, depending on the grade. Hardened and ground shafting is made to very close tolerances. Stainless-steel shafting is available, especially in small diameters.

Flexible shafting is made by tightly winding spring wire in layers. (S. S. White, Stow)

Splined shafts are made with linear ball bearings between the splines and the slider. (Warner, THK)

Threaded shafts are discussed in Chap. 7, "Linear Motion." (A screw and nut convert rotary to linear motion; so they belong in both categories.)

6.6 Clutches and Brakes

The only difference between a clutch and a brake is that a clutch couples two rotating members and a brake couples a rotating member to a nonrotating member, i.e., ground.

The variety of combinations of torque-generating effects and control effects is so great that I will list these effects separately and leave it to you to research your desired combination and size in the catalogs.

Combinations of clutch and brake in one assembly often save space, alignment effort, and cost. (Warner)

All clutches and brakes which slip under load during part of their operation time convert slip power to heat and *get hot*. The adequacy of a clutch or brake in any application depends partly on the peak torque it can transmit, partly on its rate of wear, and *primarily on how hot it gets from the slip power*. Some clutches and brakes are air-cooled with air circulated by the spinning of the parts, some have separate blowers, and some are water-cooled. Be sure that the *worst* duty cycle will not heat the device to a destructive temperature.

Continuous slip clutches are a simple form of variable-speed drive. Eddy-current, hysteresis, viscous-drag, and fluid-drive clutches do not wear and can be used to reduce the speed of an input motor or engine to a lower-speed output, in fact, to a steady torque down to zero speed. *But please remember that the heating is slip speed times torque*; at full torque and zero output speed (slip equals 1) the full rated power of the motor is converted to heat inside the clutch.

Some fluid-drive clutches with more than two rotors are torque converters, acting as continuously variable ratio gearboxes.

Continuous slip brakes are similar to continuous slip clutches and are used to maintain a controllable drag load or to prevent a load from running away. A special case is the "retarder," used on heavy logging trucks in mountainous country, which dissipates the gravity energy of a heavy load being transported downhill.

6.6.1 Torque-generating effects

Torque between the controlled parts is generated by the following:

6.6.1.1 Dry friction. Dry friction is produced by pressure between disk and disk, disk and brake pad, or drum and brake shoe.

Facing materials usually consist of combinations of steel, cast iron, sintered bronze, and composites. Materials which have been used include wood, cork, and leather.

One-way clutches ("sprag," overrunning, freewheeling) transmit torque in one direction and instantly disengage in the other direction, although they use friction and not positive engagement. There is a kind of one-way dry-friction clutch and brake which uses a helical spring that closes inward against a metal drum. These devices do not slip in the driving direction; so they do not heat up.

6.6.1.2 Lubricated friction. Driving and driven disks are alternated and stacked on internal and external splines. The stack is enclosed, and the enclosure is filled with oil.

6.6.1.3 Hydrodynamic forces. Hydrodynamic forces are present between two rotors and a circulating liquid ("fluid drives").

6.6.1.4 Viscous drag. The fluid may be either a constant-viscosity liquid with control by varying the geometry or an electrically controllable viscosity fluid.

With constant-viscosity liquids the clutch geometry may be controlled by varying the wetted area or by varying the space between the driving and driven members. (Actually, I know of no commercial variable-wetted-area clutches, but I once made a number of them in which controlled air pressure balanced the centrifugal pressure of the oil in the gap between a driving and a driven disk; changing air pressure changed the wetted area and thereby changed the torque-versus-slip characteristic. It worked very well.)

6.6.1.5 Magnetic particle. A controllable-viscosity fluid is iron particles in either oil or graphite powder. A magnetic field from a control coil makes the particles adhere and thus increases the viscosity. The magnetic-particle clutch (or brake) consists of two disks or drums separated by this fluid, the whole being part of a magnetic circuit energized by a coil. (Eaton)

Although it is not yet commercial, you should know of the electrorheological clutch (or brake). It is analogous to the magnetic-particle clutch except that the fluid changes its viscosity in an electro-static field instead of an electro-magnetic field. The limitations at the time of writing are that the viscosity changes are rather small and the electric voltages required are rather large, but electro-rheological fluids are under continuing development. (Lord Manufacturing)

6.6.1.6 Eddy-current drag. One member includes an iron core and exciting coil, and the other member includes one or more copper or aluminum disks which lie within the magnetic field of the iron core. Relative motion induces eddy currents in the disks, which react with the magnetic field to produce torque. Slip heat is generated by the electric eddy currents in the disks. Variable-speed drive assemblies are made by combining a constant-speed induction motor with an eddy-current clutch in a single package. (Eaton)

6.6.1.7 Hysteresis drag. Hysteresis drag resembles eddy-current drag except that, instead of conductive disks, there is a member made of permanent magnet material, and that, instead of eddy currents being induced, the permanent magnet material is repeatedly remagnetized, first in one direction and then in the other. The drag force is from the magnetic attraction between the iron core and the permanent magnet material. Slip heat is generated as hysteresis loss in the permanent magnet material. (Magtrol)

6.6.1.8 Positive engagement. Jaw clutches have V-shaped, square-shaped, or sawtooth-shaped teeth facing each other on the two members. They are engaged and disengaged either at zero speed or, in overload and single-revolution clutches, suddenly at running speed. Since these devices do not slip, they do not heat up. (Electric Clutch and Brake)

6.6.1.9 Electric generators and motors. Generators feeding motors are used to transmit power between prime mover and load and between load and ground in a variety of ways.

6.6.2 Control effects

Clutches and brakes are engaged and disengaged by the following effects. In most cases a spring is provided which either engages or disengages in opposition to the control effect. For example, safety brakes are typically engaged by a spring and disengaged by the control effect so that in case of a failure of control power the brake will automatically be set. In a car, the clutch is engaged by a spring and disengaged by control power, your leg.

6.6.2.1 Electricity. Electromagnets attract friction disks together, produce the magnetic fields in eddy-current, hysteresis, and magnetic-particle clutches and brakes, and are the solenoids which disengage safety brakes. Except for the on-off solenoids, electromagnets are proportional devices whose effects vary with the magnitude of the electric current.

Electromagnets operate pneumatic and hydraulic valves, either on-off or proportional, to control air- and oil-powered clutches and brakes.

If the electromagnet coil rotates, it must be fed through slip rings, a source of cost, maintenance, and unreliability. In some clutches the slip rings are omitted, and an air gap in the magnetic circuit provides a magnetic equivalent to a slip ring. The cost is a little more current and another part.

Small electric currents in the field windings of generators and motors control their torque and power levels.

6.6.2.2 Compressed air. Air is used to apply engagement force through cylinders and pistons, flexible diaphragms and cups, and expanding bladders and tubes. In rotating systems the air may pass through a rotary joint; such joints are commercial products (Deublin). Most truck and train brakes are pneumatically powered. (Horton, Eaton)

6.6.2.3 Hydraulics. Hydraulic pressure in a cylinder can change either the displacement or the pressure of the clutch or brake moving parts. Most automobile brakes are hydraulic.

6.6.2.4 Centrifugal force. There is a class of clutches whose sole job is to engage when an induction motor comes up to speed. Most induction motors produce little torque until they are nearly at full speed, so they cannot start against loads with high starting torques. These clutches use centrifugal force produced by the *motor* speed to couple the load to the motor and automatically decouple the load when the motor stops. No other control devices are needed. Usually the coupling is to a pulley or sprocket coaxial with the clutch, so a neat package is mounted on the motor shaft.

The torque effects used include dry friction, liquid coupling between rotors with buckets, and dry metal particles centrifuged between corrugated rotors. (Zurn)

6.6.2.5 Torque. Overload clutches release or just slip if torque exceeds a preset level. Release overload clutches transmit the torque through a spring (plus a cam or lever), and when the spring exceeds its set force, it permits a positive-engagement clutch to disengage. Typically a manual reset is required. This kind of clutch is analogous to an electrical circuit breaker, just as a shear pin is analogous to an electrical fuse. A cheap overload clutch is a pair of friction disks loaded by a spring and adjusting nut. The clutch slips on momentary peak loads, but it burns up on continuous overload (which may be the lesser of two evils).

6.6.2.6 Angular position. Single-revolution clutches are engaged and disengaged by a cam or lever. When a momentary signal causes engagement, the load rotates one revolution until it reaches its initial position, at which time the clutch disengages.

Hilliard catalogs a variety of unusual clutches.

6.6.2.7 Human. Levers operated by human hands or feet operate many clutches and brakes.

6.7 Rotation Transmission
6.7.1 Shaft couplings

Shafts are joined end to end by flexible couplings and by universal joints.

U joints (Fig. 3.33) require that the centerlines be free to intersect (the U joint will constrain them to do so) but may have substantial angular misalignment and furthermore will couple without looseness other than that of the U joint bearings. The angle between the shafts introduces a harmonic into the relative rotation, but two U joints, properly phased, between misaligned but parallel shafts as in Fig. 2.2 cancel each other's error.

Flexure versions of the two U joints and central shaft set in a single package are made by Thomas and by Renbrandt (Fig. 3.16). They are made in sizes from instrument to very-large-power. They have zero backlash, which is a particularly valuable feature in reversible servo systems.

Flexible couplings such as Fig. 3.17 permit both linear and angular misalignment to some degree, but much less so than the arrangements of Figs. 3.16 and 2.2. A metal bellows is used as a zero backlash and single-moving-part flexible coupling in the instrument range. Elastomeric couplings are made from small through very large sizes. They have the advantage of damping torsional oscillation. Other flexible couplings use chain, gears, and a variety of other elements.

The Oldham coupling transmits motion and torque with one sliding part.

6.7.2 Gears

Gears are so extensively treated in textbooks and handbooks that I will give no descriptive matter. However I will name Boston Gear and Browning as publishers of instructional catalogs and vendors of a large variety of stock gears and gear assemblies. For instrument-size-gear catalogs I suggest Berg, Sterling, and Stock Drive Products. Each carries a large variety of instrument-size motion transmission products.

6.7.3 Gearless speed reducers

1. Harmonic Drive,* a large-ratio coaxial transmission using a flexible member whose teeth operate as inclined planes against a ring of stationary teeth.

2. Cycloidal roller drives. Several companies make large-ratio coaxial drives in a large range of torque ratings. (Wood's Sons, Dolan-Jenner Industries)

6.7.4 Friction drives

Belts and pulleys are made in a tremendous range of sizes and materials. Among these are the following:

6.7.4.1 V belts.
V belts are made in many cross sections and many lengths. Matching single- and multiple-groove pulleys are standard (Gates). There is a segmented belt which can be threaded around its pulleys, which enables the use of a V-belt drive where the belt cannot be slipped over the pulley end.

There is a class of variable-speed drives which uses a V belt between two pulleys. The two sides of each pulley are separate, and their spacing is adjustable. The net effect is to make a pair of pulleys with adjustable pitch diameters and a corresponding adjustment in speed ratio.

6.7.4.2 Multi-V belts.
Poly V† belts have a large number of small V

---

*Harmonic Drive is a trademark of Harmonic Drive Division, Quincy Technologies.

†Poly V is a trademark of Goodyear Tire & Rubber Company and of Browning Manufacturing Division, Emerson Electric Co.

ribs side by side to get the wedging friction of a V belt with the flexibility of a flat belt.

6.7.4.3 Flat belts. Flat belts are made in many sizes and of many materials, including woven fibers, fiber-elastomer combinations, plastic films, and metals. Some are used as conveyor belts. Seamless flat belts are either woven as large-diameter tubing, which is then cut into short lengths, or cut from flat stock and stretched into belt form (Tilton). Continuous belting stock is cut to suit and spliced with commercial splices (Clipper). Belting for conveyors is made with a variety of surfaces, e.g., high-friction, low-friction, chemical-resistant, sanitary, metal, etc. (Baldwin)

Pulleys are either crowned or flanged to prevent the belt from drifting off.

Flat belts can be run at very high speed.

6.7.4.4 Tooth belts. Tooth belts ("timing belts") are flat belts with molded teeth, usually on one face, and steel or fiber tension reinforcement. They work with matching sprockets. Tooth belts have the positive-drive properties of chain and the quietness and high-speed properties of belts. They need no lubrication. They are made in sizes from instrument to large-power. They are manufactured by power transmission companies. (Wood's Sons)

6.7.4.5 Round belts. Round belts are used for low power. They can flex in every direction and so are suitable for intricate drives and for high speeds. O rings are often used as round belts. High-strength adhesives make it possible to make seamless round belts from continuous stock.

A long helical spring with its ends hooked together is used as a round belt to drive movie-projector rewind reels. It stretches and slips under load and thus can provide a constant-torque drive.

6.7.5 Chains

Chain-and-sprocket drives use chains ranging from tiny bead chain to enormous high-power and high-tension roller chain. For high speeds there are silent chains.

Roller chain is useful for converting rotary to linear motion in a variety of conveyors. Conveyor chains are made with a variety of links made for the attachments of fixtures and with large-diameter rollers which can support long flights of chain along straight tracks (Diamond). Efson makes linear chain support rails of wear-resistant plastic.

6.7.6 Indexing drives

Gears, belts, and chains deliver motion in constant ratio to their input motion. There is a class of transmissions which transfer motion with a variable, typically an interrupted, ratio. For example, a variety of linkage and cam drives do this.

A class of drives has been developed using nonuniform screw threads or cams to drive a circle of cam followers on the output. Most are designed for rather large torques and are of heavy construction. They are intended primarily for dial tables and linear chain carriers in automatic-transfer manufacturing machines. (Commercial Cam, Ferguson)

A classic drive is the Geneva motion which produces an intermittent-output rotation from a continuous-input rotation. Commercial Geneva drives are available both as complete packages and as independent parts. The disadvantage of the Geneva drive is that the peak output velocity is high. (Tangen)

A number of companies make standard two-motion part transfer mechanisms with a single-shaft input. The output motions typically consist of a lift combined with either a swing or a linear carry. These mechanisms are intended for automatic manufacturing service. The phrase "pick and place" is used to describe their action. (Sankyo America)

6.7.7 Variable-speed drives

6.7.7.1 Variable-speed motors. Electric motors and controls provide a variety of variable-speed drives. Some are dissipative, and some are not. Some provide dynamic braking, and some provide regenerative braking. Controllable electric motors, particularly with the advent of solid-state electronic controls, are the most adaptable to most stationary applications.

At one time, controllable-speed motors were mostly dc motors with commutators, which were a source of maintenance effort. Now controllable-speed motors are made without commutators, using either an electronic equivalent or variable-frequency ac.

6.7.7.2 Motor and slip clutch. A fixed-speed motor, usually induction, and a controllable slip clutch such as an eddy-current clutch (Eaton) make a variable-speed drive. However, this system dissipates substantial power below top output speed. For loads whose torque decreases with decreasing speed, such as blowers, this disadvantage may not be significant.

6.7.7.3 Friction. Many variable-speed drives have been made by using friction between hard rotating parts with a range of diameters (such as a cone) to get the equivalent of an infinitely variable gearbox. Most have been superseded by electrical variable-speed drives.

6.7.7.4 V belt. Variable-ratio V-belt drives are described above under "V Belts." Some high-power drives of this sort use belts made of links with cross slides which engage pulley grooves. (Reeves, Electric Clutch and Brake)

Internal-combustion engines with fuel throttles are variable-speed drives. Steam engines with steam throttles are variable-speed drives.

6.7.7.5 Hydraulic. Hydraulic systems based on variable-displacement pumps and fixed-displacement motors are used for high-power drives and some high-power servos. (Parker Hannifin)

6.7.7.6 Gearshift. Automatic-gearshift systems, particularly planetary drives with friction brakes on different rotating portions, are widely used in automobiles. Manual gearshifts, with clutches, are much simpler if they are acceptable.

Gearshift systems have high efficiencies at all ratios.

Chapter 7

Linear Motion

7.1 Bearings, Wheels, and Tracks

7.1.1 Roller and track matched sets

(Tsubakimoto, Thomson Industries, THK)

7.1.1.1 Round tracks. Round tracks and recirculating ball bearings are illustrated in Figs. 3.35 and 3.36. Round tracks and recirculating chained rollers are illustrated in Fig. 3.44. For short-distance travel the track is supported only at its ends, and the bearing completely encircles the track. For long-distance travel the track is supported along its full length, and the bearing has a gap to clear the support. The tracks are hardened and ground to close tolerances.

Some bearings have self-aligning ball races (Fig. 3.34) (Thomson), and some have fixed-axis races and require accurately aligned supporting holes for the bearing but have other advantages (Barden). Pillow blocks with the ball bearings built in are available if it is desirable not to machine accurate bores.

Tracks are made from $1/8$ in to 4 in in diameter in carbon steel and stainless steel and in lengths up to 14 ft. Tracks and bearings are also made in metric sizes.

The same round tracks support chained roller blocks (Fig. 3.44) (Thomson). Two or more twin chained roller sets in a V array on one track and one or more single sets on the second track constitute a semi-MinCD system. If there are only two twin sets and one single set, the system is a fully MinCD tripod; the chained rollers themselves self-align with the tracks. Chained roller systems are capable of carrying very large loads. The chained roller system is intended for horizontal tracks only since it requires a seating force to maintain contact between rollers and track. The ball-bearing systems enfold the track and need no seating force; so they can be used in any orientation, even with the load hanging from the track.

Thomson makes adjustable track supports which enable each track to be screw-adjusted vertically and horizontally.

7.1.1.2 Nonround tracks. Balls rolling on round tracks have high contact stresses, which limit their load capacities. Therefore systems have been developed in which balls roll in grooves that fit like the grooves in rotating bearings and have much more load capacity per ball.

The tracks in such systems are usually rectangular instead of round and have two or more ball grooves. The sliders usually are not round and have two or more recirculating ball races. If the load torques are not high, a single track and slider can be a complete matched-set (see Sec. 7.1.2), linear-motion suspension. On the other hand, if the load requires two tracks, the system is overconstrained unless lateral freedoms are provided for one of the tracks or between the load and one of the sliders.

Transportation "monorails" are not monorails. Typically they have two track surfaces on each side of a monolithic rail and one track surface on top, making up a pentarail rather than a monorail. However, the single solid concrete rail is better looking than the framework of a conventional rail pair and its supports.

7.1.1.3 Roller-bearing systems. Matched sets are made with pairs of tracks separated by rollers. There are sets with a single pair of rolling surfaces and sets with two pairs, providing yaw constraint. (Schneeberger)

Recirculating roller sets which you can place between your own flat ways are available. (Tychoway, Schneeberger)

7.1.2 Complete matched sets

A complete matched set contains a base having two parallel tracks and a slide having matching bearings. All assembly and alignment are done in the vendor's factory, the design is usually RedCD, and the accuracy of many such matched sets is very high. The phrase "linear slide" is generally used to describe such products. (Linear Industries, Anorad)

The "nonround track" systems described above are a special case in which both tracks are integral with the base and both bearings are integral with the slide.

Two-axis matched sets are made in which two such sets are stacked at right angles. The phrase "X-Y table" is generally used to describe such products.

Modular sets are made which can be stacked to suit your needs. In the precision instrument regime, Klinger catalogs a large variety of linear- and rotary-motion and other modules to use in building systems.

You may add your own controls and drives to a linear slide or an X-Y table, or you can buy them already furnished with motors, feedback transducers, and electronic control circuitry. (Anorad)

All the above deals with rolling-bearing products. For machine tool service you can still buy slides with dovetail and rectangular ways and with acme drive screws. (Milwaukee Slide & Spindle)

7.2 Wheels

In the roller and track systems described above the load is carried on balls and rollers without axles. The balls and rollers support load on opposite ends of their *diameters*, and they move at *half* the speed of the supported loads. This half speed either imposes the need for a recirculating system or else limits the travel. Some time ago this inconvenience was solved by the invention of the "wheel," which supports load at opposite ends of its *radius* and which moves at the *same* speed as the supported load to which it is attached by a shaft and rotary bearing. The recirculating system is replaced by the shaft and bearing. There is no limit to travel distance other than the length of the track or road. The need for a system to constrain motion to the desired path still remains.

7.2.1 Wheels for flat paths

An enormous variety of wheel sizes and types are made as commercial products, most of which are well known. Therefore, this section will merely give a listing to remind you of what you can get. (Albion)

Wheels are sold with integral bearings, either sleeve, ball, or roller, and sometimes with integral shafts in the assembly.

Wheels are made of metal or plastic, either one-piece or built-up.

Wheel rims may be integral with the bodies or may have tires. Tires may be steel to resist abrasion or of elastomer to yield over small bumps and to cushion the ride. Elastomeric tires may be solid or be inflated with air for additional softness. Polyurethane solid tires resist creep, i.e., do not develop flat spots if left stationary under load for a long time. Inflated tires may have treads to diminish skidding on wet ground.

Very soft wheels are made for conveyors to support sensitive products. (Fairlane, Stilson)

Cam followers are relatively small, accurately made wheels of hardened steel with integral needle bearings. Their original intent was to follow the surface of a cam as their track, transmitting very heavy loads. They are useful in machines as guide rollers and as load-bearing wheels. They are made with both cylindrical and crowned rolling surfaces. Cam followers are made by bearing companies. (Torrington)

7.3 Wheel Steering

Steel wheel rims may be formed to steer the wheel along a track. Railroad wheels are tapered and fixed in pairs to common axles. They steer themselves in the direction of having the same rolling diameter on their tracks, a soft centering constraint. They also have flanges as safety caging constraints if the taper steering is momentarily insufficient. Some steel wheels have twin flanges to constrain them on their tracks.

Caster assemblies cause their wheels to turn into the direction in which the load is moved rather than using wheel steering to establish that direction. Casters are made in many sizes, with vertical axle or flange mounting, with and without ball bearings on the wheel axis and on the caster axis, with and without locking brakes, with and without suspension springs, with matching noncastered wheels, and with a great variety of wheels. (Albion)

7.4 Wheel and Track Matched Sets

Accurate complete matched sets are made which consist of steel wheels with V grooves rolling on steel tracks having a matching V edge. They are made in single-axis and two-axis (X, Y) configurations. (Bishop-Wisecarver, RolerTrac Division of Master Machine Tools)

Nonaccurate matched sets are made by the user by combining commercial steel wheels having a 90° groove and commercial angle iron laid on the ground with its right angle up.

Nonrolling plastic bearings for sliding along rectangular rails are made by Precision Laminations.

7.5 Hydrostatic Sliding Bearings

Pneumatic hydrostatic bearings are made which work against smooth concrete floors. They support very large loads which can then be moved about by hand. (Airfloat)

Ceramic ways with pneumatic hydrostatic bearings are made by National Machine Systems.

7.6 Lead Screws and Nuts

Screws and nuts convert rotary motion to linear motion. In this section we deal with screws and nuts for generating linear motion in machines, not as fasteners. In machines they are called lead screws and nuts. (Warner, THK)

Acme threads are still made. For jacks, vises, and clamps their high friction has the advantage of unidirectional locking: the load cannot force the screw to turn. Jack assemblies are made which incorporate a rotating nut on bearings in a gearbox and a worm and gear to turn the nut; the screw does not turn. (Duff-Norton)

Ball-bearing screws and nuts provide recirculating balls between the nut and the screw, with corresponding round-bottom threads. It is common to provide twin nuts with an antibacklash spring. (See Fig. 3.24 for a MinCD antibacklash design and Chap. 16 for a general discussion of backlash.)

The low friction in ball screws makes it possible to have very steep threads without binding. It also makes it possible to rotate the screw by pushing the nut. Figure 14.5 shows a robot gripper which is operated in that way.

Linear actuator assemblies are made as integrated packages. They include a ball screw and nut, motor, controls such as limit switches and motion transducers, an output slide which looks like a piston rod, and attachments such as rod ends at each end. (Industrial Devices)

7.7 Belts, Chains, and Ropes

Belts and chains are like screws and nuts and like racks and pinions in that they convert rotary motion to linear motion. They are discussed in Chap. 6, "Rotary Motion," because they are also used to couple one rotary motion to another rotary motion. (Boston)

When you design a system in which rotary motion is converted to linear motion by belt, chain, or rope, be sure to consider the elasticity of the material as it affects the dynamics of your system.

Ropes are made of wire, plastic, and natural fibers. Very-small-size wire rope, pulleys, and end fittings are made by Sava.

Rope and pulleys can provide a mechanical advantage, as described in elementary physics books.

Chapter

8

Power

If your product has moving parts and is not just a static structure, you will require some form of power to make the parts move. You may also need power for light, heat, sensors, control, and other nonmotion uses. This chapter reviews the commercially available forms of power and the devices for providing and transmitting them. Chapter 15 has a broader discussion of power.

Actuators and controls convert an energy source into mechanical force and work in a mechanism and the devices which turn them on and off or otherwise adjust their output. All the devices described below are commercially available components.

8.1 Available Forms of Power

Electricity   Flywheels
Hydraulics   Heat engines
Pneumatics   Fuel burners
Explosives   Human
Springs

8.2 Power Sources

8.2.1 Electricity

Electricity is drawn from utility power lines or is produced locally by engine generators, hydroelectric turbines, windmills, primary batteries, or solar cells. Human-powered generators are made for emergency service.

Devices for conversion between ac and dc, voltage changing, safety, and sensing and indicating, switchgear, and wiring, and storage batteries for absorbing and returning electricity are all made commercially and in profusion. All these devices are commercially available in a large range of sizes and types by many companies and rarely need special engineering unless quantities needed are very large—or you are in the electrical device business.

8.2.1.1 Motors. In size, motors are made from subminiature, as used in cameras, to thousands of horsepower.

In environmental resistance, motors are made for ordinary indoor service and for waterproof, explosionproof, military, aircraft, food industry, chemical industry, and many other services.

In types of electricity used, motors operate on dc, on ac, on voltages from 3 V to thousands of volts, on voltage directly from the power source, and on voltages from both simple and complex controllers. Most are for dc or 60-Hz ac supplies, but commercial motors are made for 400-Hz aircraft power and for a variety of other supplies.

In speed, motors are made for exactly constant speed (ac synchronous), approximately constant speed (induction, the cheapest and most widely used), adjustable speed (dc shunt), and speed which varies with load torque (dc series). The most common speed is approximately 1800 r/min, but standard motors are made with many speeds. It is common practice to make motor assemblies with integral gearboxes so that one purchased assembly has a low output speed and a correspondingly high torque. (Bodine)

In position, stepping motors move in proportion to the number of electrical input pulses.

Servomotors, with feedback devices and control amplifiers, can be made to do almost anything you ask, although not necessarily cheaply. Unless you have a very capable electrical engineer aboard, you should get the servomotor, feedback devices, and amplifier from the same manufacturer and work very closely with the manufacturer's engineers in defining the specifications for both its drive and your load.

Electro-Craft publishes a valuable text on dc motors and servos. I have found Industrial Drives Division of Kollmorgen a responsive and helpful source of precision ac servos. Minarik and Boston Gear each catalog a wide range of motors and gear-motor assemblies. The industry is very, very large, and there are many, many manufacturers of both common and unusual motors.

Electric motors are easy to customize in many ways without much tooling cost, and most manufacturers are interested in doing so.

Linear electric motors are much in the news, but they are mostly specials at the time this book is being written and are therefore outside its scope.

8.2.1.2 Other electric actuators.
Solenoids are made for dc and for ac power and in a wide variety of forces and strokes. Most are for straight-line motion (Detroit Coil), but some rotary solenoids (Lucas Ledex) are made.

Voice coils (so called after their use in loudspeakers) can be made in a very wide range of sizes, but most are special. Voice coils are fast and are easily and accurately controllable. They are made in sizes from tiny loudspeakers to multi-thousand-pound shakers. BEI is a very capable source of special voice coils, with which I have worked.

Polarized armature actuators are short-stroke, reversible-force devices, somewhat similar to voice coils in principle, which are used as servovalve actuators and polarized relay actuators.

Piezoelectric crystals can be used as mechanical actuators where short stroke and low force are permissible and very-high-frequency response is desirable. The crystals are commercial, but I know of no commercial actuators.

High-power spark discharges from a capacitor bank are equivalent to explosive actuators. They are used as the propellant in a Star Wars rail gun, for example. Similar discharges into a coil can cause electromagnetic repulsion forces large enough to cold-flow metal in a manufacturing process (Maxwell Laboratories). The electrical components are commercial (except for the metal-forming coils, which are custom), but I know of no standard commercial actuating devices.

8.2.1.3 Heaters.
Electric heaters do not produce motion but are widely used in many kinds of machines; so it seems appropriate to refer to them as output devices. They transfer heat by conduction, convection, and radiation. (Watlow)

8.2.1.4 Electrical controls.
Commercial electrical control devices for the actuators listed above include the following:

- Relays (electro-mechanical and solid-state)
- Switches and pushbuttons, limit switches
- Fixed and variable resistors (potentiometers, rheostats)
- Circuit breakers, fuses, and motor starters
- Indicating lamps and meters

   (Allen Bradley, Cutler-Hammer)

Electronic controls for speed, position, torque, etc., are mentioned above.

Sequence controls such as programmable controllers and machine tool NC are described in Sec. 9.10.

8.2.1.5 Wiring devices. Electrical power and control devices must be interconnected with wire of the correct sizes, with adequate insulation, adequate protection from being damaged, and adequate protection from causing a fire.

The commercial components are:

Wire   Connectors
Conduit   Markers
Cabinets   Ties
Terminal strips

(Allen Bradley, Square D, Minarik, Cutler-Hammer)

Special devices are made to handle cables to moving portions of a large machine. These include hollow chains, flexible ducts, reels, and festoons, all commercial products. (See Sec. 3.5.1.7.) (Gleason)

8.2.2 Hydraulics

Most hydraulic power is produced by pumps driven by electric motors. Some is produced by pumps driven by engines, often as an auxiliary load on a vehicle propulsion engine. Some is produced by a human operating a reciprocating lever on a piston and cylinder (e.g., an auto jack or hydraulic tool), and some is produced from air power via an air cylinder driving a hydraulic cylinder to get a small volume of high-pressure oil.

A large variety of hydraulic fluids is available commercially.

Hydraulic power supplies comprise some combination of sumps, filters, pumps, motors, regulators, accumulators, valves of many kinds, gauges, and connectors, all joined by tubing and support structure. All these components are available commercially from many manufacturers (Parker Hannifin). There are contract manufacturers that make hydraulic power supply assemblies to specification.

8:2.2.1 Cylinders. Fluid-pressure cylinders to generate linear motion and force are the principal hydraulic and pneumatic actuators. Commercial cylinders are single-acting and double-acting, have piston rods emerging from one end or both ends, may have cushioning chambers at the ends of travel, may incorporate flow control valves in the cylinder heads, may be designed for maintenance of seals or for discard when worn, and have a variety of mounting features and other special features. Standard commercial cylinders are made in all stroke lengths from fractions of an inch to many feet and in diameters from a fraction of an inch (Clippard) to approximately a foot.

"Rodless" pneumatic cylinders are made with a flexible wire rope instead of a rigid piston rod or a magnetic or mechanical coupling to an external slide alongside the internal piston. (Tol-O-Matic)

Cylinders are made with electrical control valves built into the cylinder heads.

A unique cylinder having zero leakage and negligible friction uses a flexible cup to contain the fluid instead of sliding seals (Bellofram). Short-stroke cylinders for valve and truck brake actuation use flexible diaphragms. An elastomeric sack into which compressed air is fed serves as an actuator or spring or suspension. (Firestone)

Rotary cylinders exist in several forms. In some, a single vane rotates in a cylindrical chamber with different pressures on each side. In others an assembly is manufactured in which a pair of cylinders is coupled by a rack and pinion to a gear on the output shaft. In another, a linear piston is pushed along a rod with a long-pitch screw thread coupling the piston and the rod so that the linear motion of the piston rotates the rod.

8.2.2.2 Motors. Hydraulic and pneumatic (Aro) motors continuously rotate their output shafts just as electric motors do. They are made in sizes from fractional horsepower to integral horsepower. Some hydraulic motors are made to have very slow speed but enormous torque, without gears. Because of their different fluid compressibilities, hydraulic motors can be accurately controlled in speed, while pneumatic motor speed varies widely with load torque.

8.2.2.3 Control devices. Directional control valves are made in many sizes. There are spool valves, slide valves, poppet valves, and others. Flow configurations vary from two-way to five-way. (Parker)

Directional valves are operated by hand levers, mechanism motion, electrical solenoids, fluid pressure in pilot cylinders, and direction of flow (check valves).

Throttling valves adjust output pressure or flow and respond to the output pressure or flow acting on the control orifices or to a manual adjustment as in a faucet. Safety valves respond to *input* pressure.

Servovalves are throttling directional-control valves, usually with an electrical input. (Schenck)

8.2.2.4 Plumbing. Fluid power components are interconnected with:

Steel tubing   Fittings
Copper tubing   Manifolds
Flexible hose

All these are commercial products in many sizes and types except for manifolds. Manifolds are usually special except for those made by valve manufacturers to couple a plurality of valves without fittings and tubing.

The fluid power industry is very large, and there are many manufacturers. Parker publishes a series of texts and teaching catalogs on fluid power.

8.2.3 Pneumatics

Most factories have a central air compressor from which compressed air is piped around as a local utility. Usually the pressure varies between 80 and 100 lb/in^2 and the air is dirty, wet, and oily. Any machine intended to use such air should be provided with a filter, pressure regulator, and lubricator at the air entrance point. Many companies make such FRL sets as standard modular combinations, and the output air is quite suitable for air cylinders (Norgren). When the pneumatic devices in the machine are sensitive, you can buy coalescing filters which will remove all traces of oil, refrigerating desiccators which will remove all traces of moisture, and high-precision pressure regulators. (Fairchild)

Air compressors, other than factory utility machines, vary from tiny machines for artists' airbrushes through portable paint-spray machines to engine-driven machines to supply construction project jackhammers (Gast, Ingersoll-Rand). All are commercial.

A very convenient air supply for portable tools is a tank of compressed air or carbon dioxide. The tank must be carried around, but no long air hose need be dragged around.

Both hydraulic and pneumatic power supplies must be routinely maintained. Filters must be replaced in both, and lubricating oil must be added to pneumatic systems. Hydraulic systems are unforgiving: a little bit of dirt, and they will fail. Pneumatic systems will deteriorate but still operate for a long time if they are neglected. Therefore, some factories with poor maintenance discipline neglect pneumatic systems and then complain bitterly of their "unreliability."

Pneumatic pressure is usually less than 100 lb/in^2, and hydraulic pressure is usually above 1000 lb/in^2. Therefore, hydraulic pumps and other devices are usually much smaller than pneumatic devices of the same power or force.

Air is quite elastic, and hydraulic fluids are almost inelastic. There fore, the control properties are very different and may determine which to choose for each application. Remember the general principle that the fewer the energy forms in any machine, the better for cost, complexity, and maintenance.

Pneumatic power is less efficient than hydraulic power because the energy of compressing the air is usually dissipated when the air expands freely at the exhaust. On the other hand, pneumatic components and power supplies are usually less expensive. Both generally use an electric motor or an engine as a prime mover; so the energy efficiency of using power directly from the prime mover is higher. Usually, however, energy efficiency is much less important than the other properties of the energy form.

8.2.4 Explosives

The usual euphemism is "powder-actuated devices," which sounds less dangerous. Such a device delivers very high power for a short time and in a small package. It must be reloaded between uses or discarded after use.

Powder in blank cartridges is used in tools for stud driving. Powder is used as a source of stored pneumatic power for the cylinders of bolt or cable cutters in space vehicles. Long linear-shaped charges are used to cut large holes in sheet metal. In manufacturing, there are several processes which use explosives for laminating, forming, and bonding. Usually the powder charge is electrically fired. Several companies are listed in the directories as makers of powder-actuated devices.

Most powder-actuated devices are cylinders similar to pneumatic cylinders except that the powder charge is built in (or loaded in as in a gun) and ignited either electrically or by a firing pin. The design requirements are so unusual that these devices are made only by companies which specialize in them. I once made a cartridge-powered bolt cutter; it burst, and the cartridge case almost killed one of us, but it missed. They do have lots of pep in small packages, though.

8.2.5 Springs

Springs are discussed in Sec. 9.6. There is a large industry which manufactures both stock and special springs.

Springs, like storage batteries and flywheels, store and return power in its original form instead of converting it from some other form. Energy storage springs to drive mechanisms are usually flat spirals and are made commercially by spring companies. Associated Spring publishes teaching catalogs on many kinds of spring. The biggest historical use of spring power was in clocks and watches.

8.2.6 Flywheels

Flywheels are rarely used except to smooth out the portions of an engine cycle. However, they are used as an energy source in friction welding and in brake testing. Experiments and calculations have been made to use modern high-tensile materials in flywheels to hold large amounts of energy.

8.2.7 Heat engines

Internal-combustion engines are made in sizes from tiny engines for model airplanes through fractional horsepower for hand-carried tools to very large multiple-horsepower machines for ship propulsion. They burn compressed gas for low pollution, gasoline, and diesel oil.

Steam engines were once made in many sizes but are now made only in large sizes except for models, hobbies, and experimental work.

If wind and rain be considered large-scale heat engines (powered by solar heat), this would be a good place to list windmills and water turbines as commercial prime-mover components.

8.2.8 Fuel burning

Fuel is burned for heat power in process industries. (At one time there were quick heaters for automobile interiors in winter which burned gasoline directly without waiting for the engine to heat up. I cannot find an explanation of why this is no longer done.) There are commercial burners of many sorts to feed, carburet, and support combustion. Welding and burning torches are special cases.

8.2.9 Human power

Human power is used in a multitude of devices including turning a can opener, loading paper into a computer printer, steering a lawn mower or chain saw, propelling a bicycle or a human-powered airplane, hoisting and sheeting sails on a large racing yacht, lugging hand baggage down miles of airport corridors, and swinging the weights of a self-winding wristwatch. Some of it is unskilled, such as turning a hand crank or foot pedal; and some is highly skilled, such as cranking an orthopedist's hand drill.

Human power as a substitute for motors or engines is rarely useful except in sport or in emergencies when the machines fail. Emergency radios are sometimes powered by human-driven generators. It is also economically useful for intermittent low-power applications or when control skill is required, as in manufacturing assembly, loading paper into a machine, opening doors, and the like.

Bicycle pedals get more useful power out of a human than any other device, particularly if there is a speed-matching transmission, as in a 10-speed bicycle. The human-powered airplanes use bicycle drives coupled to their propellers. I have suggested that yacht racers use fixed bicycles instead of hand cranks to drive sail-handling ropes. (Treadmills are almost as good as pedals and were used to power the hoists that built the Gothic cathedrals, but they are bulky.)

I have resisted the temptation to identify animals as a source of power, although in the history of technology they were once the major source. I know of nothing you are likely to design for the western world which could benefit from animal power except, perhaps, an improved arctic dogsled or racing sulky.

Chapter 9
Other Components

9.1 Semifinished Materials

The easy way to start a design is with ideas and raw metal. If you are designing a product to be made in huge quantities, this is the right thing to do. If your quantities are limited, then you can save large amounts of engineering, manufacturing, and maintenance cost by using commercial semifinished components and materials for parts of your product.

Material manufacturers have taken advantage of commonalities which appear in many designs to offer semifinished products and materials. These save you a great deal of tooling and manufacturing cost if you take the trouble to learn about them before you design. Conversely, they encourage you to design things you would not undertake if you had to start from common materials. Here is a partial list.

Shafting is made finished to close tolerances, finishes, and hardness.

Mechanical tubing for turned parts is made with a wide range of ID, OD, and alloy. You may easily design many parts to corresponding dimensions to minimize metal scrap.

Tooling plate of aluminum, steel, and magnesium is made with no internal stresses, with ground surfaces, and in many thicknesses. Although intended for making jigs and fixtures, it can be machined into other products.

Unusual rolled and extruded shapes of aluminum are stocked by many metal warehouses. The aluminum companies, such as Alcoa, catalog these shapes.

Extruded plastics are made in a tremendous variety of standard shapes and materials.

Perforated metal sheet in a tremendous variety of patterns is cataloged by Harrington & King.

Clad metals are made in which a decorative surface metal covers a steel or other base metal.

Laminated plastics are made in which a decorative surface is laminated to a strong core.

Honeycomb is made in a wide variety of core and surface materials.

Wood paneling is made in which a decorative surface ply is laminated to an inexpensive core.

Drawn steel wire and rod are available in a variety of cross sections including gear teeth ("pinion rod").

This section has by no means listed every semifinished material which may be useful to you, but it should encourage you to collect catalogs on those classes of semifinished materials which you might use.

9.2 Structural Systems

The principal approaches to modular frameworks are tubing with joint fittings, formed sheet metal with joint fittings, and rolled and extruded sections.

Structural tubing is made in a wide variety of diameters. Some joint fittings are designed to be attractive, as for railings. (RK Industries, MK, Weldun)

Formed sheet-metal channel comes in a wide range of sizes and with attachment fittings. (Unistrut)

Rolled-steel sections and a very wide selection of aluminum extrusions are stocked by metal warehouses and are cataloged by steel and aluminum companies.

9.3 Enclosures

Commercial enclosures range from tiny to huge and from rough and functional units to attractive objects suitable for your product's exterior.

Standard electrical equipment cabinets are made to meet each NEMA specification. (Hoffman)

Some companies also make a line of internal and external parts for terminal strips, fascia panels, and the like.

Krause-Hinds catalogs explosionproof enclosures.

Zero catalogs a very large range of drawn aluminum boxes and other enclosures.

A number of companies make lines of electronic enclosures based on standard relay rack and panel dimensions. (The word "relay" here is a carryover from olden days when these racks were used for telephone relays.) (Bud)

Large electronic distributor catalogs have sections on cabinets and enclosures and are a good starting point to scan the market. (Newark, Arrow Electronics)

9.4 Machine Modules

In the last section we saw a variety of semifinished materials which you can use to advantage in making frameworks, enclosures, and parts. In this section we will list mechanisms and structures which can be completed to form a variety of products, perhaps including yours.

Machine bases have a tabletop of flat ground steel, ready for machining the mountings of machine parts, and mounted on legs. They come in a very wide range of sizes.

Dial tables rotate a table (dial) in discrete steps for transfer manufacturing machines (machining or assembly). They are typically both robust and accurate, and their incrementing mechanism would be extremely expensive to make in small quantities. Major manufacturers are Ferguson and Commercial Cam.

Linear transfer tables are linear analogs of rotary dial tables.

Modular assembly machines are made by Bodine and others. They include modular transfer machines and modules for performing a variety of operations at the different stations.

Modular robot axes are made by PHD, Mack, and others. They allow you to build up a particular configuration of motions to suit your needs.

Pick-and-place mechanisms for feeding and unloading machines are made by several companies. Whether or not they are "robots" is a pedantic question but may be a sales feature for the manufacturers.

Part feeders, with or without tooling for your particular parts, are made by Detroit Screwdriver, FMC-Syntron, and others.

Drilling and milling heads, complete with guideways and feed and retraction mechanisms, are standard products.

Precision spindles for rotating grinding wheels or other tools are made by Pope and others.

Machine slide modules are made by Milwaukee and others.

Multiaxis motion machines in modular form are made by Klinger for instrument-size work and by Anorad and others, complete with drives and electronic controls if you wish.

Only a large directory such as *Thomas* has a complete list. I suggest that you guess category names for what you hope you might find and then study a directory to see what you can buy instead of designing and making.

9.5 Fasteners

9.5.1 Threaded fasteners

Since most products contain two or more parts held together by fasteners, the variety of fasteners which have been developed is large, and the number of fastener manufacturers is large, and the number of standards is large. It is rare that a special fastener must be designed because a standard one cannot be used.

As a starter, I recommend the following references and catalogs:

Federal Standard Screw Thread Standards for Federal Services, FED-STD-H28 (formerly *National Bureau of Standards Handbook H28*), U.S. Government Printing Office, Washington, D.C.

Machinery's Handbook, Ref. 29. This is indispensable as a data book on sizes and types of fasteners of all sorts as well as many other mechanical components.

See the catalog of Illinois Tool Works and the Southco Inc. *Fastener Handbook*.

9.5.1.1 Data and specifications.
The basic elements of threaded fasteners which you must consider are:

- Diameter
- Length
- Thread pitch
- Material
- Surface finish
- Grade
- Length of thread
- Head forms (bolt and nut)
- Point and thread form, if it is to be self-threading in metal, plastic, or wood

9.5.1.2 Threaded inserts.
Nuts have been developed which can be fastened into holes in soft metals, plastics, and wood to provide the equivalent of tapped holes in strong material. Heli-Coil catalogs a line of spring-formed thread reinforcements. Others are solid and are held in place by cold forming, by ultrasonic heat forming in plastic, and by insertion from the back side.

9.5.1.3 Sheet-metal nuts.
Very inexpensive nuts are made of hardened sheet metal. They tend to be self-locking and, when used with hardened screws, are surprisingly strong. (Tinnerman)

9.5.1.4 Other forms.
A variety of fasteners and other devices are made to provide a tapped hole or threaded stud.

Weld nuts and studs are made to be spot-welded (Ohio Weld Nut), or arc-welded (Nelson Stud Welding) to a member, or staked (Pem) into a hole in a soft metal member.

Firestone makes a combination nut and cold rivet which permits riveting two members together and then adding a third with a screw.

Several kinds of screws are made with a smooth portion under the head and a threaded portion at the end. The smooth portion serves as a dowel to align the assembly more accurately and with greater shear strength than do screw threads.

Some screws and nuts are made with lockwashers permanently retained as a subassembly. (Illinois Tool Works)

A very useful form of screw is headed with a hardened spike which is driven into concrete or steel with a powder-actuated gun. The result is a stud fixed to the concrete or steel base.

9.5.2 Thread locking

Since threaded fasteners tend to loosen under vibration, expansion and contraction, shock, etc., many products have been developed to prevent loosening.

9.5.2.1 Lockwashers. Lockwashers cut into the fastener and a fastened part to produce high friction between them in the loosening direction and to provide a short stiff spring. Lockwashers are made as split rings having a single tooth on each face and as multitooth rings with internal teeth or external teeth or both and sometimes with integral electrical lugs. Some screws and nuts are made with lockwashers permanently retained in grooves. (Illinois Tool Works)

A common design error is to put a smooth washer between the lockwasher and the fastened part to prevent abrasion of the fastened part. This destroys the effectiveness of the lockwasher since it is precisely that abrasion which does the locking. The lockwasher now locks the fastener only to the smooth washer, and both are free to turn against the fastened part.

Remember that if there are a screw and a nut, *both* must be prevented from loosening.

A variety of products have been designed to be more effective than lockwashers and to eliminate abrasion from lockwashers.

9.5.2.2 Locknuts. A second nut, called a locknut, under the main nut increases thread friction.

9.5.2.3 Lock wire. The screw heads and nuts are cross-drilled, and a steel wire is passed through the hole, twisted, and fastened to a part to be fastened.

9.5.2.4 Castellated nuts. The nut is slotted, and the bolt is cross-drilled. A cotter pin is passed through a pair of slots and the drilled hole and prevents rotation of the nut relative to the bolt in either direction.

9.5.2.5 Insert nuts. An unthreaded plastic insert is staked into the end of the nut. When the bolt is forced through the insert, it forms threads which retain an elastic grip on the bolt threads through many revolutions. (Elastic Stop Nut)

9.5.2.6 Insert bolts. For use with tapped holes, a plastic strip is inserted into the bolt and acts as does the insert in the insert nut.

9.5.2.7 Deformed nuts. Part of the nut is bent out of its nominal geometry and is elastically bent back by the threading in of the bolt. The resulting friction resists loosening. The nut may be reused many times.

9.5.2.8 Adhesives. A drop of certain adhesives, which harden only when trapped in a thin film, bonds the screw threads together. The adhesives are made in different grades of strength to require different loosening torques, and even when the bond is broken, the friction drag remains quite high. The surfaces must be clean to start with and may require a primer. (Loctite)

My own practice in designing machines I really want to stay together is to favor deformed nuts first and thread adhesive second, to require that the assembler tighten every fastener with a torque wrench and put a colored dot on it to say that it has been done, and to require that an inspector put a differently colored dot on it to indicate that the connection has been tested with a torque wrench. I prefer nuts to tapped holes because a stripped thread does not mean a spoiled part and the quality control of the commercial nut thread is more reliable than the quality control of a tapped hole.

9.5.3 Nonthreaded fasteners

9.5.3.1 Rivets. These fasteners depend on hot or cold flow of part of the fastener to retain it. Among these are:

- Solid rivets
- Drilled rivets
- Tubular rivets
- Rivets with a draw pin to deform the head
- Swaged ring (pseudo nuts swaged onto pseudo screws)

Drilled rivet ends are often incorporated into electrical terminals and spacers.

Combination rivets and nuts provide a fixed nut mounted on a rivet.

9.5.3.2 Other fasteners. Other fasteners, some of which have been described elsewhere in this section, are:

- Retaining rings
- Dowels (smooth, deformed, tapered)
- Good old-fashioned cotter pins and snap-in variations of them, still very much alive

9.5.3.3 Latches. Innumerable latches, quarter-turn fasteners, knobs, and the like are on the market. (Southco, Panel Products, Reid)

9.6 Vibration and Shock Absorbers

Reread Sec. 3.4, "Soft Constraints."

Vibration and shock are attenuated by a number of commercial components. Some are used to mount vibration- and shock-*sensitive* assemblies such as radios and shock-*generating* assemblies such as engines and punch presses. Others are used to decelerate ("cushion") moving bodies at the ends of their strokes.

9.6.1 Shock mounts

"Shock mounts," which are both shock- and vibration-attenuating mounts, interpose a dissipative elastic member of elastomer, cork, wire mesh, or wire rope between the mounted body and ground. In a car or truck suspension the two functions are usually divided between an elastic spring for support and a dashpot ("shock absorber") for energy dissipation. Compressed air can be used for suspension and its flow through an orifice for dissipation.

Machinery shock mounts are used under punch presses and other shock- and vibration-*generating* machines to attenuate transmission to ground. Some use cork and elastomers and also serve to anchor the machines (Unisorb), and some use air bags (Firestone).

Most shock mounts are passive devices, but Lord has developed electrically controllable orifices in liquid-filled mounts.

Lord and Barry provide teaching catalogs on shock mounts.

9.6.2 Shock absorbers

Shock absorbers are dashpots which force oil or air through an orifice to dissipate the energy of an impact. The orifice may be varied by the position of the plunger so that a more uniform braking force is generated.

Teaching catalogs on oil-filled shock absorbers are published by Ace and on air-filled shock absorbers by Airpot.

9.7 Springs

Springs are pervasive in all kinds of mechanisms; so it is not surprising that there is an industry which manufactures and inventories a large variety of stock springs. The types of springs which are commercially available include:

Helical compression	Belleville
Helical tension	Gas
Spiral (clock spring)	Liquid
Constant force	Elastomer
Wavy washer	

Associated Spring and its divisions publish teaching catalogs of spring design as well as lists of stock sizes. Lee publishes a catalog of many sizes.

Gas Spring Corp. makes springs using trapped gas in compression instead of stressed metal.

Liquid-filled springs are quite stiff; they are mostly used in dies.

Epoxy laminated with glass fibers makes flat springs with very long fatigue life; 3M makes laminates for the purpose, but you must cut out the shape to suit yourself.

The only elastomer springs I know of which are stock products are the multitude of shock mounts, bumpers, etc., which are cataloged by rubber-molding companies. However, these companies will mold your designs to order. In general, elastomer springs have the greatest energy storage per unit of volume when they are stressed in shear. Lord has made very large cylindrical springs loaded in torsion.

Wire-spring-manufacturing machines are easily programmed, so it is relatively inexpensive to buy special springs.

9.8 Lubrication

Commercial lubrication devices range from oil cups and oil drip reservoirs (Gits) to metering systems which send measured quantities of lubricant through separate tubes to individual bearings (Trabon, Lincoln) and oil-spray systems (Alemite).

Lubricants are available based on petroleum, whale oil (still used for some instruments), soap (usually to hold oil in the form of a grease), silicones, fluorine-based synthetics, and dispersions of graphite and molybdenum disulfide powders. The many lubricant companies (many are chemical companies like Hooker, 3M, Monsanto, and Du Pont) and all the oil companies (Exxon, Shell, etc.) provide teaching catalogs.

Dry lubrication is available both as loose powders of graphite (Morganite) and molybdenum disulfide and in adhering coatings containing those powders or using silicones or fluorocarbons.

Bearings which have a low coefficient of friction and can run without lubricant are made of graphite and a variety of plastics.

As a design note, vibration is a friction reducer or eliminator. In testing reciprocating-engine airplanes it was common practice to make movies of the instrument panel. When this was tried with early jet planes, the lack of vibration let the instruments' hysteresis introduce errors. Shakers were added to the instrument panel to restore the instruments' accuracy. (Now electronic transducers are used.)

Another technique of reducing friction in one direction is to drive the parts along each other in the transverse direction. Stiction goes to zero. For example, a complex linkage to prevent "window locking" in robot insertion tooling was developed at MIT. A major Detroit automation company bought a license. The company people set up a demonstration in which the visitor was challenged to insert, by hand, a ball bearing into a close-fitting hole. After he failed, the demonstration mechanism put it right in. When my turn came, I rotated the outer race as I pressed, and the bearing went right in. They threw me out.

Surface treatments (chemical conversion coatings, diffusion coatings) to minimize friction are available from a number of companies. These include phosphate coatings for steel and hard anodizing plus the addition of plastic lubricants for aluminum. (General Magnaplate)

Hard surfacing to resist abrasion without lubrication is provided by hard anodizing on aluminum and by sprayed coatings of hard alloys or ceramics applied by flame spray or plasma arc spray.

9.9 Seals and Guards

Commercial seals to exclude dirt and retain pressure and lubricant for stationary, linear sliding, and slowly rotating service are available in the following forms:

- O ring (with round or other sections) [Parker (handbook), Precision Rubber Products (handbook), Minnesota Rubber]
- Flat gasket
- U ring
- V ring
- Spiral and helical

Seals are made of elastomers, fibers, and even metal.

Seals can be molded to metal washers or other supports. (Parker)

Adhesive seals for nonsliding service, i.e., sealants, are also made of viscous liquids which harden in place. Some are made of silicones (Dow Corning), some of asphalt, and some of hardening adhesives (Loctite).

Commercial seals for high-speed rotary service are spring-loaded face seals (Advanced Products). A special case is the rotary union for conducting fluids into and out of rotating machines. (Deublin)

Flexible bellows and sliding covers which protect machine ways and the like are available commercially. (Gortite)

9.10 Sensors and Displays

In addition to their use in laboratories, many commercially available sensors and displays are included in products for sale and for special machines for in-house use. (Omega, Omron)

A sensor responds to the quantity measured and either displays it to an observer, as in a voltmeter, or feeds it back to a controller, as in a thermostat, or both. A "transducer" is a sensor.

"Part present," or "proximity," sensors use electrical, electromagnetic, optical (i.e., photoelectric), or ultrasonic effects. Optical systems range from a simple photocell responding to an interrupted light beam to geometrical gauging systems to computer analysis of a television image.

A display may be a mechanical displacement, such as a pressure gauge needle, or an electrical display such as illuminated digits or a cathode-ray-tube pattern, or merely a warning lamp.

Sensors are also used in automatic feedback control systems. Feedback can be mechanical, as in an automobile choke responding to a heat-sensing bimetal or an old-fashioned steam engine governor, or pneumatic as in process industry pneumatic control systems, or a fluidic control, or electrical, as is now most common. Electrical systems may be analog or digital.

9.10.1 Parameters

Commercial sensors are available for the following parameters:

- Temperature
- Pressure
- Displacement (linear and angular)
- Acceleration (linear and angular)
- Weight
- Liquid flow rate
- Gas flow rate
- Voltage
- Current
- Light intensity (total and spectrum analyzing)
- Viscosity
- Chemical composition

Other quantities may be converted to these and then measured; for example, the volume of liquid in a tank may support a float whose displacement is measured and the display calibrated in gallons.

For any parameter a sensor will usually be limited to some range and to some environmental limitations.

I know of no sensor which measures stress, although one would be very useful. I do not mean a system which measures *change* in stress, as does a displacement transducer ("strain gauge"), but one which can be applied to a steel beam already under a static load and measure its stress. Consider this an invitation to invent.

There are single-use sensors such as pyrometric cones for high temperatures, stress-recording lacquers which develop crack patterns corresponding to the stresses in the underlying solid, and magnetic-powder and fluorescent-dye surface and subsurface flaw detection procedures. (Other nondestructive testing techniques include ultrasonics, x-ray, and several nuclear effects.)

9.11 Sequence Controls

Many machines perform a sequence of operations. There are several kinds of commercial devices which are used to turn on and off the steps in this sequence in accordance with rules established by the designer. Sequence control is different from magnitude control, in which, usually within one step of the sequence, the magnitude of a parameter such as temperature or position is controlled in accordance with other rules established by the designer.

9.11.1 Timers

A timer controls the duration of a one-step sequence. Timers are made mechanically and electrically operated. Their output is usually an electrical switch. Multiple timers are often included in other controllers described below.

9.11.2 Drum controllers

A drum controller has a rotor on which is a set of cams or other flags. As the rotor turns, the flags operate switches to turn on and off actions in the machine. The rotor may be turned by a constant-speed motor, as in a domestic washing machine, or it may be turned in steps in respond to action completion signals in the machine. Most switches are electrical, but some drum controllers operate pneumatic or hydraulic valves directly. (Candy, Electro Cam, Furnas)

A camshaft directly driving the required motions is a classic and highly sophisticated drum controller because it controls position, speed, acceleration, and jerk of each of its outputs and does so with high accuracy and reliability.

9.11.3 Relay circuits

Relays may be wired to be the equivalent of a drum controller but with many additional functions such as interlocks and switching between one subprogram and another subprogram, depending on conditions of the machine. Industrial relays for this purpose are made by Cutler-Hammer and others. Relay circuits may include timers.

Electronic relays and modular circuits are made to be combined in an equivalent way, although they have been largely superseded by programmable controllers and computers, as described below.

9.11.4 Programmable controllers (PLC)

A programmable controller (programmable logic controller, or PLC) is a microcomputer with built-in software which allows it to be programmed in ways very similar to designing a relay circuit. The common phrase is "ladder logic" because industrial relay circuit diagrams look like stepladders. The benefits of PLC over relays are as follows:

A PLC is the electronic equivalent of hundreds or thousands of relays in a small package. It is programmed from a keyboard instead of by wiring. The condition of each "relay" is displayed on a cathode-ray tube; so debugging is easier. Many functions other than relay logic can be had, including timers, simple arithmetic, and servo controllers. There is no wearing out of relays or faults due to dirty contacts. PLCs are usually modular; so you can combine the components needed for a particular machine.

You need no knowledge of electronics or of computer programming to select, apply, and program a PLC.

There are many manufacturers who provide teaching catalogs, courses, and helpful sales engineers. I suggest, to start, Allen Bradley and General Electric.

9.11.5 Computers

Digital computers programmed in a variety of languages and in many sizes, depending on machine size and complexity, are used as machine controls. IBM makes an industrial version of a personal computer for industrial control.

9.11.6 Nonelectrical controllers

Digital fluidics is used in a relaylike manner as machine controls. The advantages are noise immunity and silence, temperature and radiation insensitivity, and absence of danger of initiating explosions.

Pneumatic and hydraulic systems use the equivalent of pneumatic and hydraulic relays in sequence controls. They are used only in relatively simple sequences, but they have the advantage in simplicity that the actuators and controllers all use the same power medium. All the major valve companies make such controls. (Note that in the pneumatic controls business the word "relay" is often used to mean the pneumatic equivalent of an electronic "amplifier.")

There are drum controllers which operate pneumatic and hydraulic valves.

Camshafts, discussed above, are nonelectrical controllers.

Punched-tape and punched-card controllers are mostly obsolescent, replaced by magnetic or semiconductor computer memories, but you should know about them so that if you should have a special case in which they would be appropriate, you can use them. The player piano uses a multicolumn punched tape about 1 ft wide in which the holes are sensed by vacuum and the vacuum operates small pneumatic cylinders. Teletype tape, sensed by mechanical pins or by airflow through the holes, was the original input to numerical control systems for machine tools. To this day, Jacquard looms weave decorative fabrics under the control of punched tapes (in the form of wide punched cards hinged together to form a tape in which one hole controls one thread crossing).

Also obsolescent are patchboard electrical memories in which a program is established by wires plugged into an array of connectors.

9.12 Tooling Components

Although the design of jigs and fixtures is outside the scope of this book, many of the standard components made for this service are useful for other mechanisms. These include ball feet, adjustment screws, wheels and knobs, tooling balls, spherical washers, etc. (Carr Lane)

9.13 Permanent Magnets

Permanent magnets are used as latches and other devices. (Thomas & Skinner)

9.14 Lamps

Available lamps include tiny indicators, incandescent illumination lamps of sizes from fractional watt to thousands of watts, small high-intensity halogen, sodium arc high-power, and fluorescent. There are sockets and fixtures in great profusion for all classes of lamps. (General Electric)

9.15 Nameplates

Nameplates are made by:

- Silk screening
- Die casting
- Etching

- Engraving
- Printing

Separate nameplates are made of metal, plastic, and film. They are fastened to the product by fasteners or adhesives. Integral nameplates are made as part of the product fabrication.

9.16 Pumps and Blowers

Commercial pumps and blowers are made with output pressures ranging from high-vacuum to at least 15,000 lb/in^2. They handle gases, liquids, and slurries. Some work by positive displacement, some by centrifugal force, some by venturi effect, and some by electro-magnetic force. They handle flows from milliliters per hour to thousands of gallons per minute. Some are sealed by rubbing seals and some by hermetic magnetic couplings. Some resist internal and external corrosion, and some do not contaminate fluids, including human blood. No one company makes all types, but the directory categories are easy to identify.

9.17 Miscellaneous

There is a great variety of specialty hardware and other unusual products which can save your company much design time and manufacturing cost. For example, Boker makes an enormous variety of washers. It is a good idea to read the ads and collect catalogs of products which you might want someday. When you are doing a design, it can be well worthwhile to do a speculative search of directory categories to see what is out there in both major components and small items such as brackets and clamps.

Exercises in Design with Commercial Components

For each of the designs you have made in the previous exercises using MinCD, semi-MinCD, and RedCD, find and specify as many commercial components as you can. First make up a set of specifications for size, power, speed, torque, duty cycle, etc. Then consult directories for vendors, secure catalogs, select specific components, and justify your selections with calculations.

Part 3

Topics in Design Engineering

Chapter 10

Designing with Uncommon Manufacturing Processes

Whenever you conceive a part, you should have in mind at least one way of making it. The more ways of making parts you know, the more kinds of part you are free to conceive and the more imaginative you may be in designing your mechanisms.

Although it is desirable to design around your company's own plant and machinery, there may be significant benefit in using a process you don't have in house. If so, you should propose contracting the part to a specialized vendor. The number, variety, and capabilities of such vendors is breathtaking.

Even the largest companies contract a substantial portion of their work to vendors, either to get capabilities not in house, or to get more capacity, or to get lower costs. Most parts are subject to a "make or buy" decision.

Many companies manufacture *nothing* and contract out everything except engineering, marketing, and finance. I once ran a small mail-order business making and selling drafting and engineering instruments I had invented. In a complicated manufacturing cycle each vendor did its work and drop-shipped its output to the next vendor in the process. I just made paper in my home.

Some processes are suitable only for mass production; others are valuable in making models and short runs, either for small-quantity production or for testing designs prior to mass production. Your knowledge of the economics of tooling and setup is as important as your knowledge of the capability of the process.

The following is a list of manufacturing processes applicable to the manufacture of mechanisms. Read the list and ask yourself if you understand each one and if you think it may help with your present design problems. The more of them you know, the better you are as a mechanism designer.

Casting: sand, permanent mold, investment, die

Plastic molding

Powder metallurgy

Flame cutting, including plasma arc

Flame spraying, including plasma arc

Machining (small, large, precise, etc.)

Electrical-discharge machining (EDM)

Electro-forming

Chemical milling

Chemical etching (through thin sheets)

Explosive forming

Explosive laminating

Diffusion bonding

Magnetic-pulse swaging

Tube bending

Press working: (punching, forming, drawing, coining, fineblanking, hydroforming, rubber-die cutting and forming, short-run)

Cold heading

Upsetting

Forging

Extrusion: plastic, metal; impact extrusion

Die drawing of shapes

Printed-circuit-board etching, drilling, laminating

Electrical cables (wire combinations)

Electrical cable harnesses

Electrical control panels (industrial, quantities of one and up)

Engineering ("job shops")

Nameplates: engraved, etched, cast, silk-screened

Heat treating

Electron-beam welding

Laser welding

Laser cutting

Water-jet cutting

Ultrasonic welding

Friction welding

Double-disk grinding

Shot peening

Blanchard grinding

Laminating (plastics, metals)

Vapor deposition

Ultrasonic welding

Nondestructive testing

Roll forming

Roll-formed strip

Spinning

Isostatic pressing

Vacuum deposition

Vacuum impregnation

Surface finishing: corrosion resisting, abrasion resisting, low-friction, decorative

Reference

Bralla, James G. (ed.): *Handbook of Product Design for Manufacturing*, McGraw-Hill, New York, 1986.

Chapter 11

Manufacturing Engineering

Manufacturing is a badly neglected art, and the United States is suffering badly from this neglect.

Manufacturing engineering is treated only superficially in engineering schools; I know of none that gives a degree in manufacturing engineering. (You can get a degree in industrial engineering, but that is not a field in which you design machines.)

In industry, manufacturing engineering has low prestige and low pay. I was told of one large company whose manufacturing engineers were unsuccessful product engineers who trickled down to the bottom of the organization, which is where manufacturing engineering ranked.

The title of "manufacturing engineer" is commonly conferred on workers who have no college degree but have worked their way up from manufacturing or drafting jobs. In product engineering such people are considered technicians and have such titles as "senior designer." (I am not a snob. I have much more respect for such achievers than for college graduates with low energy and a weak sense of responsibility. But no amount of practical experience is equivalent to a college education in mathematics, science, and analytic engineering.)

11.1 What Is Manufacturing Engineering?

It includes the design of the following:

11.1.1.1 Standard machines. You design standard and modified manufacturing machines for sale to using manufacturers. This is a class of product design in which the products are manufacturing machines. You need a knowledge of manufacturing technology to be able to design the machines to be cost-effective producers.

11.1.1.2 Special machines for sale. You design special manufacturing machines within a special machine company, for sale under contract to using manufacturers. Professionally, you live the life of a product engineer except that there is more pressure on you to meet cost and completion goals and you are limited to small-lot fabricating techniques.

11.1.1.3 Special machines for your company. You design special machines for your own factory. This work is done, typically, under your manufacturing engineering management, and now the problems start.

11.1.1.4 R&D. You perform manufacturing technology R&D, development of processes, and selection of purchased standard or special equipment. This work can be as creative and satisfying as any engineering can be, but it is also done under your manufacturing engineering management.

11.1.1.5 Tool design. You design dies, molds, jigs, and fixtures. Much of this is routine in nature, applies handbook data, is nonanalytical, and is suitable for senior technicians ("designers"). In some cases an engineer can contribute great value, particularly when something really new is being developed. Some of this work is contracted to specialty companies, and some is done in house under your manufacturing engineering management.

11.1.1.6 Planning and scheduling. To a large degree this is technician-level clerical work assigned to "manufacturing engineers" and will drive a real design engineer into a depression or the classified ads. It is often done under your manufacturing engineering management.

11.1.1.7 Maintenance. This also is technician-level work. It has a very real priority (the factory must be kept going), but if maintenance is assigned to a design engineer, it destroys the continuity of the engineer's attention to real engineering jobs. It is often done under your manufacturing engineering management.

As you may have anticipated, I have some unkind things to say about much of the "manufacturing engineering management" I have encountered. Its members, too, have climbed the ladder from manufacturing technician and have a corresponding education. The climb has been made in a rough environment; so they are true bull-of-the-woods material.

The environment they grew up in and maintain is intolerant of error. A product development engineer is expected to make errors as the price of being innovative. A manufacturing engineer is punished and threatened by these people if he or she makes an error. Guess what this does to the advancement of manufacturing technology and the attractiveness of manufacturing engineering as a profession. (The zeroth rule of personnel supervision is that punishment will train anyone not to make errors. Punished persons will never again make an error. They will do little or anything else and will certainly never take a risk, but they will not make another error.)

A relatively minor discouragement of the profession is that its office facilities are often the worst available in the company.

Typical manufacturing engineering staff members are culturally coarse because of their backgrounds. This environment is not attractive to college graduates.

I have seen exceptions, of course, mostly in the electronics industry. IBM, Hewlett-Packard, and 3M are conspicuously different and better.

What to do?

The Society of Manufacturing Engineers struggles mightily to raise the education and prestige level of the field, but it does not manage our companies.

I do not pretend to the wisdom to solve the problem. The solution is the responsibility of senior executives who make much more money than I do for their presumed ability to solve it. However, I will make a few suggestions.

11.2 Suggestions

11.2.1.1 Risk responsibility. Include a risk factor, written down, in the "justification document" made out for every major purchase and project. Establish a tradeoff policy of risk against innovative benefits. Remove from the engineer who advocates the buy or performs the development the sole responsibility for the occasional loser.

11.2.1.2 Technician work; engineering work. Separate the clerical planning operation and the maintenance operation from the new equipment and R&D engineering operation. Assign good engineering managers to the engineering. (This last is a real chicken-and-egg problem.)

11.2.1.3 Motivation. Try to foster a real respect for and encouragement of innovation. Use praise, awards, and rewards. At least avoid punishment for reasonable error. (Yes, I understand the difficulty of deciding which errors are "reasonable" and which errors are the result of negligence and incompetence.)

11.2.1.4 Offices. Give manufacturing engineers (at least the real ones) the same office quality as product engineers.

11.2.1.5 Education. Encourage the colleges to train and designate "manufacturing engineers." Money is very, very encouraging. Endow a chair of manufacturing engineering. Provide scholarships and fellowships in manufacturing engineering. Attract students to become manufacturing engineers. Good salaries attract good engineers.

Another approach is to do nothing but give lip service and wait for the Japanese to take over the company and do it right—or compete it into closing down.

Chapter

12

Optimum Level of Mechanization and Automation

Mechanism designers are concerned with manufacturing mechanization and automation for two reasons. The first is that you want to design your products for the most economical manufacturability. The second is that you may design manufacturing machines.

Manufacturing is done with many degrees of mechanization, of which automation is the greatest. To condense the language, this chapter refers to all of them as "degrees of automation." To select the best degree for each use you must consider cost of the machinery, quantity and variety of the work to be done, cost and availability of workers of different skills, and several kinds of policy other than cost.

This chapter classifies degrees of automation, discusses the benefits and penalties of each class, discusses labor reduction, and identifies some applicable management and government policies.

12.1 Classification

There are many kinds of manufacturing machinery. This chapter deals mostly with handling, assembly, and metal-fabricating machines, but the principles discussed apply to all. A "machine" as used here includes the machine, its tooling, and all the work required to plan, procure, modify, and install the machine and its associated equipment to get into production. With automatic machinery the cost of this work may equal the cost of the basic machine (see Chap. 13, "Robots").

Degree of automation may be classified as follows. The classification is not absolute, and you will think of machines which do not fit this classification exactly, but it is a useful arrangement for the ideas discussed.

12.1.1 Fully automatic

This class includes:

1. Robots (jointed-arm and Cartesian) for machine loading and unloading, for tool handling (mostly spot welding), and for assembly
2. Automatic storage and retrieval systems (ASRS), a class of Cartesian robot
3. Automatic guided vehicles (AGV)
4. Automatic-assembly machines
5. Automatic programmed machine tools (NC, CNC, template, or cam-controlled)
6. Looms and spinners

Most of these machines are reprogrammable to some degree. Programming may be mechanical (e.g., cams and templates) or electrical (e.g., digital computers or relays). Once installed, such machines are reprogrammed only for variations within the task performed at the installation. For example, the ASRS and the AGV are reprogrammed every cycle, and the CNC machine tool may be reprogrammed after every part or batch of parts, all by higher-level computers in a control hierarchy.

12.1.2 Powered machines with human control

This class includes:

1. Fork trucks (These do the work of ASRS and AGV, but they have human drivers.)
2. Load balancers (These enable a human to move heavy parts into and out of machine tools and pallets. The human "drives" the balancer along three or more axes via valves or switches, and the balancer supports the weight.)
3. Cranes (bridge, jib, or other. Cranes are a form of load balancer but can be made to cover large areas and carry very heavy weights. They usually do not include pitch, roll, or yaw axes, but the flexibility of their rope permits a human to control these directly.)
4. Conveyors (Loaded and unloaded by humans or machines, these transport products horizontally.)

In all of these, the machine does the work with engine or motor power, but a human provides the control or program.

12.1.3 Combination human and automatic

In this class the actual work is divided between human and machine. In this class are:

1. Hand-loaded trucks, machine tools, etc.
2. Single-station assembly machines in which a human moves together parts which are then fastened by a powered machine such as a riveter
3. Multistation assembly machines in which certain stations use humans, other stations are automatic, and the transfer conveyor is automatic

12.1.4 Human work with power tools

Power tools in human hands require more skill and may be slower, but they cost much less than automatic or semiautomatic machinery. Planners sometimes neglect the possibility of increasing productivity with better, more powerful, and less common hand tools and choose much more expensive machines unnecessarily. Such tools are powered by hydraulics, pneumatics, and electricity. Many are semiautomatic, such as rivet feeding and setting machines. Tools exist which cut, join, fasten, and do many other things that increase the productivity of humans.

12.1.5 Human workers with special hand tools

Humans may be made more productive with purchased hand tools of kinds not usual in the particular industry and with inexpensive but cleverly designed special hand tools. For example, standard dental and surgical instruments can sometimes improve the productivity of skilled workers such as those in electronics and instrument manufacture.

I once visited a company in which a friend of mine, a physics Ph.D., worked. It was the ultimate high-tech shop making an instrument for NASA to be landed on Mars. My friend was project manager. Clean room, computers all over, the works. Quality control consisted in having manufacturing teams of two workers each, an operator and a watcher. The watcher never touched anything but never took eyes off the worker's hands. (You can also see this in Las Vegas, the "workers" being dealers and players.)

I said to my friend that the workers were using the same beat-up hand tools they probably used to fix TV sets at night, and would not the miniature work they were doing benefit from using dental and special tools? He struck himself on the head in chagrin and said, "Oh my God, of course, but the project is so close to the end that it would shake everything up if we tried that now!"

Most power machines must have special tooling, which is taken for granted, but manual workers often continue to use only common tools, often not in the best condition.

12.2 Assembly Kits

Productivity can sometimes be increased if work is divided between preparing kits and using kits. (A kit is a box of parts and materials to make one product assembly. A sophisticated kit box has a separate pocket for each part so that the assembly worker need not search for it.) Transferring parts from inventory into kits requires less skill than assembling the parts from the kit into the product. By providing the skilled assembler with a kit of parts, you eliminate the worker's need to reach into inventory and you enable many assemblers to work from a common inventory reached by only a few kit assemblers. Making kits is itself much easier to automate than is assembling products. (I have proposed such automation by using the tape feeders described in Sec. 3.5.1.6, but I have not seen it done.)

12.3 The Benefits of Automation

- Less labor cost
- More uniform product
- Less loading time for machine tools
- More safety
- Less worker boredom (better quality of life)

12.4 Justifying the Cost of Automation

In the developed countries of the world the primary motivation is less labor cost. The cost of workers varies widely, but in some factories it is very high. The morale of the workers (their willingness to work fast and well as well as their attitudes) varies from factory to factory and from department to department within a factory. High worker cost and low worker morale justify the cost of a high degree of automation.

In less developed countries the considerations are at least quantitatively different. Particularly where labor rates are low, only lower levels of automation are justified.

The duty cycle of the machinery affects justification. Two or three shifts, long runs between setups for changes, and 52-week operation tend to justify investment in automation.

The above are subject to at least approximately objective calculation. The following are not.

12.5 Policy Questions

These questions, whose answers affect automation decisions, must be decided by factory managers or by government.

1. How important is the prestige of having advanced technology in the factory? In China I was told by the manager of a research institute, "China needs robots!" The voice and body language told me that the institute and Chinese industry need prestige. They already have lots of cheap labor.

2. What is the willingness to subsidize advanced technology beyond direct economic justification in order to learn from it?

3. What amortization period should be used in calculating benefits? Should it be the same for all the above classes, or should there be an amortization period subsidy for high technology?

4. What is the willingness to risk some failures in exchange for the opportunity to achieve some successes? This willingness varies with individuals, it is an industrial cultural attitude which differs in different industries, and it varies with anticipated risks and rewards. Relevant words are "courage" and "conservatism." For example, in the United States the electronics industry is much more willing to take technology and automation risks than is the shoe industry.

5. What are the policies toward workers? (Subjects include training, discipline, job subsidy, unions, and treatment of workers displaced by automation.)

6. What are the predicted changes in product type and production quantities? Should the mechanization possibilities change the predictions?

Much knowledge, ingenuity, and judgment can be valuable in deciding the optimum level of automation and mechanization.

Chapter 13

Robots

13.1 History and Myth

There have been more hype and myth about robots than about any other field of technology except man in space. I founded and operated a robot company, I detest hype, and I am obsessed with the real world; so I think it useful to you if I describe robots as I believe they really are and are not.

The word "robot" was coined about 1920 by the Czech playwright Karel Čapek in his play *R.U.R.: Rossum's Universal Robots*. He derived the word from the Czech word for work. The play is an ordinary sci-fi opus about mad scientist Dr. Goll, who invents humanlike machines to do the world's work; they conspire to turn the tables; but the good guys win. The play is dead, but the word caught on and is now the same in all languages.

Unfortunately the myth of the artificial human stayed on too. The myth has been used to imply false exaggerations of real technology in order to sell both products and careers. A substantial fraction of the sales of the robot industry has been to corporate executives who don't know what robots are but who want to be up to date. They bought robots and put them into laboratories "to see what they could do." (In contrast, every robot that my company sold was justified as a real cost reduction in a real job in a real plant.)

I was once visited by an engineer from the Army who wanted robots to load tank cannons. I told him that robots were large general-purpose machines which would never fit into a tank turret. I suggested that since there are a lot of tanks, he develop a special-purpose automatic loader which would fit. He said, "The colonel wants a robot." I let him go; I am not interested in government contract boondoggles.

A key part of the myth is that robots have computers with "artificial intelligence," and since artificial intelligence is part of the computer science vocabulary, much of the purchasing and application of robots by customers and much of the R&D by robot manufacturers and government laboratories were put into the charge of computer scientists. These scientists knew little of real-world manufacturing but thought they did because it had to be simple compared with their computers. In Chap. 19 I will tell you of the robot company president, a computer science Ph.D., who thought that my robots could be replaced by computer chips.

The result was a lot of lost money, a lot of disillusion in management, and a highly touted growth industry which didn't grow.

13.2 Robot Reality

Are robots fake, then? Absolutely not. Let us examine what real robots are, what they do, and what their future may be.

Robots are automatic machines. Each contains two or more axes of mechanical motion: an "end-effector," or tool, which is the business end of the robot; and a control system, usually electronic.

13.2.1 End effectors

This is a classification of end effectors which will also tell you much about where robots are used.

13.2.1.1 Fabricating tools.

- Spot welders
- Arc welders
- Deburring cutters
- Paint-spray guns
- Sealant- or adhesive-dispensing nozzles

13.2.1.2 Material-handling tools.

- Part grippers for machine loading and unloading, automatic assembly, and automatic storage and retrieval
- Mail delivery trays
- Pallet fixtures for automatic guided vehicles

13.2.1.3 Sensors.

- TV cameras

- Microphones
- Radars
- Chemical detectors
- Ultrasonic sensors

13.2.1.4 Military components.

- Warheads
- Fire controls (laser illuminator, TV, radar, sonar)

13.2.1.5 Quick-change grippers.
Robots have been made to use a number of different grippers in the same installation by having fixed to the robot a *gripper gripper*, which takes one or another *part gripper* from a rack so that the robot can handle a number of different parts. The principal application is automatic assembly. This scheme is analogous to a machine tool with a tool changer and is quite feasible technically, but there are three limitations.

The first limitation of the scheme is the large amount of time required to swap grippers compared with the time it takes to do the work of each gripper.

The second limitation is the requirement, if the robot is to handle several parts, to group several part feeders close to one robot. If you have ever worked on an automatic-assembly machine, you know that the part feeders are the weak links in their chains. Part feeders jam frequently, particularly on out-of-tolerance parts. Access to part feeders is required for initial setup, to clear jams during operation, and for maintenance. A robot embedded in part feeders is a costly and exasperating machine.

The third limitation is that the gripper gripper gets in the way.

13.3 Robot Control

Robots are controlled in ways similar to those used to control machine tools.

13.3.1 Point-to-point robots

These are positioned at a series of discrete points and do their work after they have arrived and stop at each point. Spot welders and most material handlers are point-to-point machines.

Some point-to-point machines such as mail delivery robots and automatic guided vehicles follow analog-defined continuous paths between work points, usually by tracking an electrical wire or painted stripe on the floor. Military guided missiles and homing torpedoes follow paths established by their guidance systems.

13.3.2 Continuous-path robots

These do their work while they are in motion. Paint sprayers, sealant dispensers, deburrers, and arc welders are continuous-path machines. Most control systems are digital, as in numerical controls (NC) for machine tools. Paint-spray robots duplicate the motions of a human and are one of the few types in which the jointed-arm configuration is as good as or better than the Cartesian configuration.

13.3.3 Human remote control.
Some nonautomatic machines which are remotely controlled by humans are also called robots. Since the mechanical technology and much of the electrical technology are the same as in automatic machines, the distinction is somewhat pedantic and I will ignore it.

Examples of human remotely controlled robots include traveling machines used for surveillance and handling in security, military, and dangerous environments. Such robot vehicles have been made for under the sea, in the air ("drones"), in space (satellites and space probes), and on the earth's surface. Undersea robots inspect pipelines and ship hulls. An airplane with an automatic pilot is a robot switched from manual mode to automatic mode.

The space shuttle is launched and returned as a robot which is partly automatic and partly controlled from earth like any unmanned vehicle. The astronauts just go along for the ride and to do things in orbit which could more economically be done automatically or by remote control. The forthcoming space station will be more of the same. (See Chap. 17, "Hype.")

13.4 Robot Mechanisms

Automatic machines were built for years and didn't even know they were robots. Transfer machines for automatic machining and for automatic assembly are sets of synchronized robots. Some special automatic-assembly machines are now built which incorporate stand-alone standard robots electrically synchronized with other mechanisms, part feeders, and product conveyors.

Robot axes are usually positioned by electric or hydraulic motors or cylinders with feedback control. (It is common to speak of "moving the axis" instead of moving a slide or rotor *on* the axis.) A transducer senses and feeds back the actual position, and an amplifier compares it with the desired position and powers the motor or cylinder to move the axis to that position. Feedback control servomechanisms have been developed to move large inertias, at high speed, to accurate positions, with high stability.

Some simple robots use air or hydraulic cylinders which drive their moving parts against mechanical stops.

A robot is a stack of separate linear and rotary axes, one mounted piggyback on another. (An exception is a swimming or flying vehicle in which pitch, yaw, and roll control surfaces or propellers are independently mounted on the vehicle.) Many stacking sequences of linear and rotary axes have been made, including branched stacks in which two independent axes are mounted on one supporting axis.

13.4.1 Linear versus rotary axes

Here the myth of the artificial human reappears. Animals have only rotary axes. (I suppose there are fundamental biological reasons why a linear slide cannot exist as part of a living organism.) The human arm is a stack of rotary axes: shoulder, elbow, wrist, branching into fingers with joint after joint. In fact, each of these joints except the fingers is really a multiple-axis device like a set of gimbals. Most human joints are actually ball joints, each with a large number of muscles controlling pitch, roll, and yaw.

A robot, to suit the myth and to be as humanlike (anthropomorphic) as possible, must be an "arm" made as a chain of links coupled by rotary joints (rotary axes), and most are. Furthermore, to have a usefully large work envelope, it is necessary to extend the arm to include a "waist" axis.

You, as a mechanism designer, will recognize that this cantilever jointed-arm system has the following disadvantages:

13.4.1.1 Errors. The chain of cantilevers magnifies the errors in the joints.

13.4.1.2 Flexibility. The chain of cantilevers magnifies the flexibility in the joints and links.

13.4.1.3 Inertia. The moment of inertia on each joint varies with the angles of the successive joints.

13.4.1.4 Geometry computation. It is so difficult to compute the individual joint angles to generate a desired end-effector position and orientation that computer programs are required to do so.

In actual fact, most robot designers have found it necessary to incorporate one or more linear axes in their axis stack; they just don't talk about it very much. In one common case a jointed-arm robot is mounted on a long linear axis to enable it to have a useful work envelope. The robot is described as a "track-mounted robot"; the linear axis is not included in the description as one of the robot's own axes.

13.5 Cartesian Robots

A number of companies, including mine, decided that usefulness and cost were more important than mystique; so they developed Cartesian robots. The name was taken from "Cartesian coordinates." *In general, Cartesian robots use linear axes to establish position and rotary axes to establish orientation (Fig. 13.1).* If you compare the characteristics of Cartesian robots with the above list of defects of the jointed-arm robots you find:

13.5.1.1 Accuracy. The position accuracy is the accuracy of the axis drives without cantilever magnification.

13.5.1.2 Flexibility. The structural flexibility is the flexibility of the axis structure without magnification. Many axis structures can be supported at both ends; so their stiffness is quite high.

13.5.1.3 Inertia. The inertia on each axis drive is constant; it does not vary with the positions of the other axes.

13.5.1.4 Position computation. The position of the end effector is easily computed and programmed by simple Cartesian geometry.

13.5.1.5 Control. An important consequence of this ease of position programming is that a Cartesian robot can be controlled by a simple, inexpensive commercial PLC ("programmable logic controller," or "programmable controller"), while a jointed-arm robot requires a special computer similar to a multiple-axis numerical control computer for a machine tool.

The PLC makes it easy to command the robot from a higher-level system computer. For example, in automatic storage and retrieval robots, a system computer orders the robot to store or retrieve at one of the locations the system computer controls. The PLC also easily interfaces commands and responses from associated machines such as machine tools and conveyors. It is common for a system including a jointed-arm robot to require a PLC *in addition to* the robot computer.

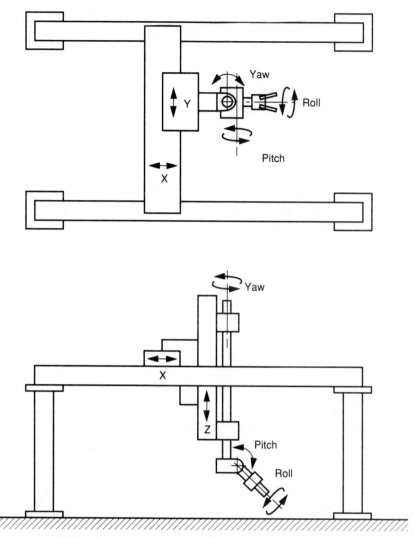

Figure 13.1 Full-bridge robot.

13.5.1.6 Modularity. The Cartesian robot system can be made modular, each axis being a separate mechanism, with standard modules of different load capacities custom-cut to the desired travel lengths.

13.5.1.7 Branching. Still another benefit of the Cartesian system is that it is possible to make branched stacks. This is a great time saver in machine loading, where one branch removes a finished part while another branch stands poised to replace it with a fresh part. Most machine tools spend a very small fraction of their time actually cutting chips, anything which reduces unloading and reloading time is a great cost reduction, and it is cost reduction which justifies machine-loading robots.

13.5.1.8 System configuration. A practical difference between jointed-arm and Cartesian robots deals with access to the served machine or process.

13.6 Safety

Every robot must be fenced and blocked. The fence prevents humans from entering the robot's motion envelope, and the blocks prevent the robot from moving outside the programmed envelope as a result of a fault.

It is common practice to install a jointed-arm robot on the floor in front of a machine tool, just where a human operator is intended to stand and enclose it in a safety fence. If the robot feeds the machine, you don't need the operator. But if the machine must be observed because of improper cutting or improper loading, a mechanic must enter the fence and stand next to the robot, which is unsafe if the robot is running. The mechanic must turn off the robot to watch the operation of the robot which he or she has just turned off. Or the mechanic takes his or her chances.

13.7 Cartesian Robot Configurations

13.7.1.1 Bridge crane. The bridge-crane configuration is shown in Fig. 13.1. It serves a working space from above. At the end of its horizontal travel it permits a mechanic to put his or her face within inches of the end effector in perfect safety.

13.7.1.2 Half bridge. The half bridge (Fig. 13.2) is intended for tasks in which the working positions lie in a vertical plane. A short axis perpendicular to that plane may be added to accommodate small variations in work positions perpendicular to the vertical plane.

13.7.1.3 Vertical bridge. The vertical-bridge configuration is shown in Fig. 2.15. Column 1 moves along track 14 in the same way that, in Fig. 13.1, the X carriage moves along its tracks. Column 1 carries a vertical slide of the type shown in Fig. 2.13, which in turn carries the axis motions suitable for the specific task, such as inserting and extracting a pallet from a shelf.

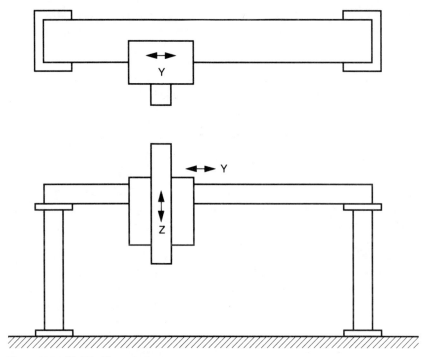

Figure 13.2 Half-bridge robot.

The myth has not yet disappeared, however. Recently an engineer acquaintance said to me, "After all, Larry, those Cartesian machines of yours are not *real* robots."

13.8 Programming

To overcome the programming difficulty of jointed-arm robots the early robot developers invented "teach mode." A set of switches or transducers is connected to a handle on the end effector, and the programmer moves the handle along the desired path. The entire robot becomes a multiaxis follow-up servo which follows the programmer's hand and records the motions in a computer memory. Not cheap, but it works. Another version of teach mode is to provide a set of jog buttons on a pendant so that the programmer can move one axis at a time until he or she gets the desired motion while the electronics records what has been done and then replays it.

There actually is one application for which teach mode is the only way to go, and that is spray painting. The only one who knows the best path for a spray nozzle, without leaving stripes and drips, is a painting craftsperson who learned that art *as* an art and who could not possibly express it in words, let alone in numbers. Therefore, to program a spray-painting robot a real spray gun is coupled to a set of transducers and given to the craftsperson. The craftsperson hand-paints a real part, his or her motions are recorded in computer memory, and the robot then reproduces the human motions.

13.9 Accuracy versus Repeatability

A machine tool which cuts a part to shape must itself be as accurate as the desired shape. On the other hand, a part-loading robot leaves no record of its work on the part (it had better not!) but must successfully load and unload the part from its machine tool workholder and from its transportation nest. Therefore the robot's position *accuracy* is unimportant, but its *close repeatability* is essential.

There is no benefit in making the straightness, squareness, and stiffness of the linear-axis ways and rotary bearings or the accuracy of the linear and rotary positioning drives compare with corresponding machine tool accuracies—providing they accurately *repeat* the positions which the commands call for.

In fact, the tolerance on robot gripper position relative to machine tool workholder position is *plus or minus zero*, since both must grip the same part at the same time during part transfer from one to the other. The brute-force solution to this problem is to let the misalignment bend the robot a little.

The elegant solution to the problem is the self-aligning gripper. It has freedoms and a small centering force which permits the gripper to align itself with the workpiece in response to small forces between the workpiece and itself. The machine tool workholder moves the gripper as the workholder engages and moves the workpiece. Thus great accuracy need not be paid for initially or maintained thereafter.

Where tolerances are too large for such passive self-alignment to work or the abutting slopes of the part and workholder are shorter than the misalignment itself, this *passive* self-alignment can be preceded by *active* self-alignment. Active self-alignment uses transducers on the robot to sense reference surfaces on the machine tool and guide the robot servos to their final position.

For example, we successfully loaded a row of four lathes with 800-lb workpieces. These fitted into diaphragm chucks with only 0.030-in clearance. The lathes were in a line extending over 140 ft of floor space, the floor was subject to settling, and the robot had to be mounted on building columns which deflected with the wind. We used both active and passive self-alignment and never had a failure.

13.10 Conversion from Task to Task

Part of the robot myth is that since the robot is controlled by a computer which can easily be reprogrammed, the robot need not be dedicated to a task but can be moved from task to task almost like a human. Thus it never becomes obsolescent. This is a perfect example of the thinking of a computer scientist unburdened by mechanical and manufacturing knowledge.

The bad news is that the application engineering to marry a robot to another machine is a major undertaking. It requires mechanical engineering to position and install the robot without interferences, provide safety fences and blocks, design new end effectors, and develop an appropriate material flow system. It requires electrical engineering for interlocks for personnel and machine safety and for combining the control circuitry for the robot and for its serviced machine or machines. Actual experience shows that robot installations are almost always dedicated and that application engineering and physical work consume a substantial part of the total installed cost.

13.11 Variation within Task

The good news is that a robot can be reprogrammed to do variations of the same task almost as easily as the computer programmers say it can. Typical variations are:

1. Loading a different part into the same machine
2. Loading components into different positions in a printed circuit assembly
3. Changing the path of a continuous-path robot

The ability to be easily reprogrammed for variations within a task is a great advantage of electronically programmed machines.

13.12 Money

With few exceptions, *manufacturing* robots do nothing that humans cannot do and are not already doing. The justification for the robots is that they do it for less. (The uniformity argument is convertible to a higher yield of good product, which is making good products for less.) I have never found a manufacturing manager interested in spending a lot of money to reduce employees' boredom. *Robots are bought as cost reductions, and they reduce cost by replacing workers for less money than the workers cost.* Typically one robot replaces one worker.

13.13 Humans versus Robots

Humans are fierce competitors of robots.

13.13.1.1 Task-to-task conversion. Humans easily transfer from task to task and accommodate to variations within task. "Reprogramming" is verbal instruction. Humans' hands are adaptable end effectors. They are not sources of danger.

13.13.1.2 Capital cost. If business declines, humans can be laid off and their costs suspended. The capital investment in robots is frozen.

13.13.1.3 Multiple tasks. A single human can perform a large number of diverse operations in the same manufacturing cycle.

13.13.1.4 Mobility. Humans can shift their workplace in a moment.

13.13.1.5 Expanded scope. Humans can fetch and deliver their own material from the stockroom when the material delivery service breaks down.

13.13.1.6 Bad-part rejection. In assembly work humans see defects and reject bad parts. And their feeders never jam.

13.13.1.7 Task modification. Humans reprogram with a few verbal instructions; so they accommodate to product and task changes with little difficulty.

13.13.1.8 Dexterity. Humans have dexterity which no machine can approach. They quickly adapt to new tools and workpieces. They have two general-purpose grippers, called hands, with multiple sensors and feedback controls, which no robot end effector can match.

13.13.1.9 Speed. In handling small parts through small distances with intricate motions, humans are much faster than robots. This capability is particularly applicable to assembly work. An exception is printed-circuit-board assembly; robots are faster because the motions are so simple.

13.13.1.10 Maintenance. If humans go down for maintenance, temporary replacements are easily found and the maintenance is done by doctors not on the payroll.

13.13.1.11 Technology. Humans do not threaten managers with a strange technology which the managers do not understand. (How well managers understand humans is a different question.)

13.14 Disadvantages of Human Workers

The disadvantages of human workers are equally well known.

13.14.1.1 Work uniformity. Human work may not be uniform.

13.14.1.2 Unions. Humans join unions which reduce the power of managers and increase labor costs. (Power ranks with religion and sex in human motivation. Money is an aspect of power.)

13.14.1.3 Fatigue. Humans suffer from fatigue, must have rest periods, and can work only 8 hours a day. Robots can work 24 hours a day without a stop.

13.14.1.4 Conflicts. Humans have personality problems which cause conflicts.

13.14.1.5 Absenteeism. Humans have absences without warning (except at the opening of fishing and hunting seasons).

13.14.1.6 Injuries. Humans have costly injuries.

All these benefits and disadvantages are well known to managers, although they cannot be quantified on a justification form and some must not be named aloud. The consequence is that the decision to automate an operation is a highly emotional one for managers, although they may insist that their decision-making process is completely rational.

13.15 Economic Justification

A common rule for justifying the investment in a cost reduction is that it pays for itself in 1 year. Therefore an *installed* robot, to be justified for single-shift work, must cost no more than the total cost of the replaced workers, usually one, for 1 year. This is very hard for a machine costing $100,000 (in 1989) plus installation. The chances for justification are very much better for two shifts and often very easy for three shifts.

There are other economic considerations. Dangerous work is expensive. (Let's leave out mere humanitarian considerations; we're dealing with money!) For sales purposes I used to equate the cost of one robot to the cost of one injured back, with no assurance that there would not be another injured back the next day.

13.16 Task Size and Force

Work which involves moving large and heavy parts through long distances is a glowing exception to the one-for-one rule. People are made in a very narrow range of strength and size. They are very fast when doing high-dexterity work moving small, light parts through small distances; they are very slow when moving large and heavy parts through long distances, even with power tools.

But there is no limit to the strength and size to which you can build a machine. Furthermore, the cost of the mechanism is less than proportional to the strength and size of its parts, and the cost of the controls is almost independent of the size of the mechanism. I therefore found, as a rule of thumb, that *big robots are justified much more often than small robots.*

13.17 Abuse Resistance

Another consideration in robot design is abuse resistance. The results of a collision by a forklift truck can be awe-inspiring.

We once built a robot for an auto company. Knowing its reputation for abusive treatment, I overdesigned the support structure to the point of being ridiculous. When the company's supervisor visited, he sneered at what he called the structure's "lightweight construction" and told me that the company was used to getting *"rugged* machinery."

13.18 The Future

The future of manufacturing robots should be one of continued development of their mechanism and control technologies, therefore of their cost efficiency, and therefore of their market. After the bad taste of the hype has left the mouths of manufacturing managers, the use of robots in manufacturing should increase for the fundamental reason of cost reduction.

Nonmanufacturing uses should be even more sensitive to increased technical capability, since fewer such uses are only cost reductions.

Chapter 14

Robot Grippers

14.1 Methods of Gripping

Material-handling robots grip their workpieces by using the following effects.

14.1.1 Friction

The most common gripper pinches the part between two jaws with surfaces which conform to the part, rather like a pair of pliers or a vise with jaws sculptured to match the part. Friction between the jaws and the part supports the part and maintains its orientation. Fragile parts may be handled by soft jaws.

Other jaw grippers are made with two, three, and more jaws. The jaws may be coupled to a common driver or may be driven independently to adapt to variations in parts and part positions. The jaws may move linearly or may pivot on axes.

If three synchronized jaws are used, the part is automatically centered in the gripper in the manner of a three-jaw chuck. Figure 14.1 shows an internal three-jaw gripper. The jaws 1 are moved by conical cam 2, operated by air cylinder 3.

Some jaw grippers are made with fixed *displacement* between open and closed positions. Compliant *pressure* is provided by elastic-jaw tooling. Some are made with fixed *force* with the engaged (closed) *position* established by contact with the part.

Fragile parts may be gripped by flexible tubes inflated with air against the part as in Fig. 3.9. We gripped an entire row of thin-wall ceramic blocks for catalytic converters between a pair of gripper tubes despite the fact that the blocks were covered with a slippery platinum slurry. There are no moving-mechanism parts in such a gripper; so it has low cost and high reliability.

A simple but effective internal-friction gripper for cylindrical parts

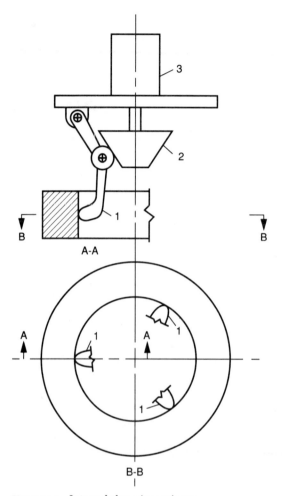

Figure 14.1 Internal three-jaw gripper.

is the twin O-ring gripper of Fig. 3.10. We mounted 10 of these grippers side by side on a bar and picked up 10 grenade bodies at a time. Again, there were no moving-mechanism parts.

14.1.2 Vacuum

Where it is usable, vacuum is the simplest technique for gripping any part. It is usable for almost any part having a flat surface on top with sufficient area compared with its weight.

The simplest vacuum gripper is a single suction cup. For heavier weights, or where the orientation of the part would be too soft with a single cup, or where a single cup cannot be placed over the center of

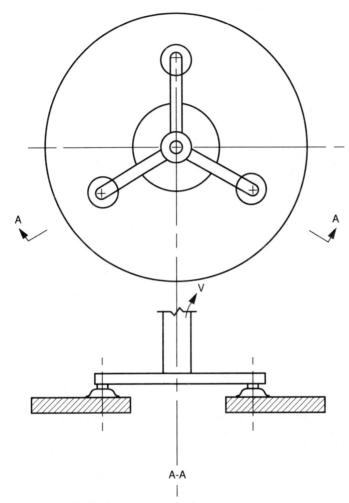

Figure 14.2 Multiple-cup vacuum gripper.

gravity of the part, multiple cups on a spider provide stability, as in Fig. 14.2.

For heavy weights or other special cases, a fitted gripper body can be made. Figure 14.3 shows a vacuum gripper for lifting heavy steel harrow disks having a central hole. An aluminum gripper disk 1 is sealed to the workpiece by two ring-shaped lip seals 2. The inner seal prevents the central hole in the part from draining the vacuum.

Figure 14.4 shows a flexible vacuum gripper for destacking large, thin sheets of printed circuit material after the stack comes from a laminating press. The vacuum grips the entire sheet area. When the gripper is raised by its suspension chains, lift is first applied to two

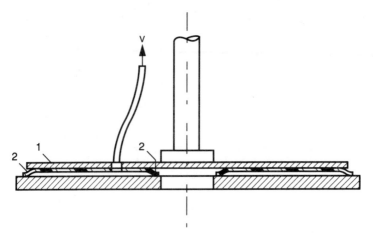

Figure 14.3 Annular vacuum gripper.

edges, which flex upward and separate the top sheet from the second, to which it tends to adhere.

Stacks of parts often adhere to each other so that when you lift the top part the second part lifts too. You can apply air blasts or mechanical separating knives to the part edges to break this adhesion.

Vacuum can be generated by a vacuum pump valved and ducted to the gripper, but the valve and tubing must be large to accommodate airflow at low pressure. A much more convenient method is to use an aspirator at the gripper and feed shop air to the aspirator from a small valve through a small-diameter tube.

Safety against vacuum failure must be considered. Some commercial aspirator grippers have an electrically released vacuum valve which holds the vacuum, if the air supply fails, until it gradually is lost through leakage.

In some cases you may be able to engage a mechanical latch as a safety cage after the part has been lifted and then release the latch by solenoid or air cylinder. In many cases the loss of air is so infrequent that a dropped part is economically permissible.

In all cases you should provide a vacuum switch to indicate the actual presence of the vacuum before lift motion can occur. (In general, in electrically programmed automatic machines, every action and required precondition should signal to the controller that it has been successfully accomplished or exists before the controller can command the next action. *Timing is not enough*. Murphy's law is always looking for an opportunity. Limit switches and other sensors are expensive, but wrecks are more expensive.)

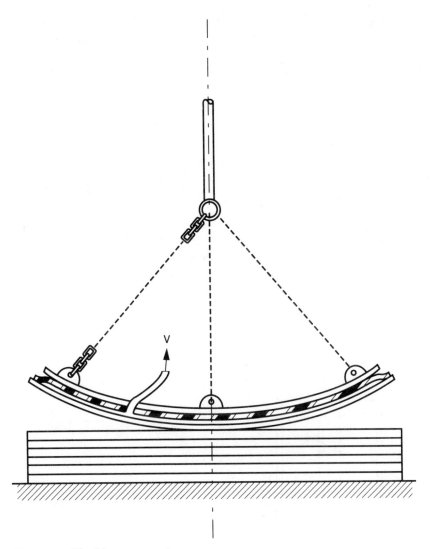

Figure 14.4 Flexible vacuum gripper.

14.1.3 Electromagnets

Electromagnets, having no moving parts, are very tempting for grippers and sometimes are the best things to use. However, they tend to leave residual magnetism in the parts gripped, and this residual magnetism collects magnetic particles; so electromagnets should rarely be used. We used electromagnets successfully to grip a set of baskets for a chemical immersion process. However, these baskets never left the process machine; we merely moved them around from one tank to another. The electromagnets and the pole pieces on the baskets were immune to the chemicals and did not contaminate the chemicals.

Grippers which must be released by solenoids or cylinders are insensitive to power failure; they "fail safe." They are also insensitive to pressure and voltage variations in power sources.

14.1.4 Special gripping devices

Parts continually appear which cannot be gripped with conventional techniques because of their shapes, surface sensitivity, workholder obstruction, environment, or some other restriction. These are a continuing challenge to your inventiveness.

For example, Fig. 14.5 shows a gripper I designed for an artillery shell 1, which is inserted through a close-fitting hole 2 into a gamma-ray beam for inspection of its high-explosive contents. Shielding required that nothing increase the hole diameter beyond the actual diameter of the shell plus a very small clearance. Gripper 3 is machined to match the ogive of the shell, then cut with helical slots 4. It is twisted by hydraulic cylinder 5, pressing ball nut 6, whose ball screw

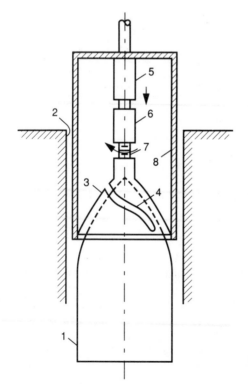

Figure 14.5 40-mm-shell gripper.

7 is attached to the gripper. Outer member 8 transmits torque to the gripper end. The gripper works great.

A simpler example (Fig. 14.6) is a gripper for an 8-in artillery shell. We were not supposed to drop it. Ever. The gripper comprises a pair of yokes 1 and gravity-engaged latches 2. The shell weight locks the latches in their engaged position, and the small disengaging cylinder 3 is too weak to operate until the shell is resting on its destination support.

An elementary but very useful form of gripper is the lifting fork, as in a forklift truck. A particular variation of the lifting fork is the lifting bed on an automatic guided vehicle (AGV), with which the AGV picks up a pallet of material that may weigh several tons. Typically the pallets engage tapered locating pins in both the AGV bed and the fixed beds to and from which the AGV transports the pallets. The pins prevent the pallet position from drifting during many transfers.

To lift blocks of soft material, grippers have been made which penetrate the material with sets of needles oriented at angles to each other. The needles hook into the material to lift it.

Sometimes a motor-driven motion within a tool eliminates the friction which prevents the performance of a task. For example, it is very

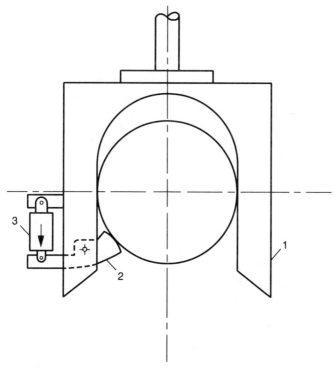

Figure 14.6 8-in-shell gripper.

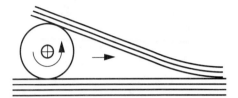

Figure 14.7 Newspaper opener.

difficult to push a cylindrical part into a close-fitting hole. However, if the plug is continually rotated by a motor, insertion is easy. We made a successful end effector in this way to put precision-ground shafts as inserts into a plastic mold in which gears were formed on the shafts. (See Sec. 9.8.)

Another example of friction "lubricated" by a motor-driven motion was the requirement to open the pages of a newspaper in an automatic insertion machine (Fig. 14.7). Every effort to insert a blade or a wedge resulted in crumpled page edges. I replaced the passive wedge with a small spinning wheel. It pulled all the pages it touched to one side, leaving the stack cleanly open for the next operation. This was one of the few times in an engineer's lifetime that a device would not go wrong: it did not matter what the wheel was made of, its speed, or its exact position; it always worked.

14.2 Environmental Limitations

Clean rooms may not have any discharge of compressed air because of dust, moisture, and oil in the compressed air. Leakage through piston-rod seals is such a discharge. Therefore, aspirator-induced-vacuum and air-cylinder-operated devices are forbidden. However, we made scavenger hoods over air-cylinder piston-rod glands and coupled them to an external vacuum pump; the construction was accepted and used in a clean room.

Grippers which must be immersed in liquids must neither contaminate the liquid nor be corroded by it.

Very hot environments degrade elastomers; so seals of elastomer are not permissible in vacuum grippers in such environments, and air cylinders and electrical insulation must be of a suitable grade for the temperature.

Nuclear radiation makes certain materials radioactive; so those materials may not be used in radiation environments.

Machine tool enclosures may be full of coolant spray and flying metal chips which stick to surfaces even when machining is stopped. Air jets may be used to clean surfaces just before gripping. In the grip-

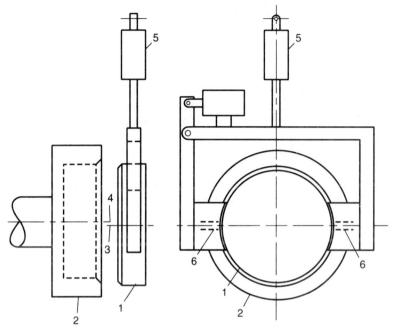

Figure 14.8 Weight-balancing gripper.

per of Fig. 14.8 air jets 6 are used as part present sensors (in the manner of an air gauge) and simultaneously clean the gripped areas of the part. (Pneumatic sensing is an old, reliable, inexpensive, but neglected technology. "Fluidic" logic technology is a relatively recent development which has not been very successful in competing with electronics except in hot, radioactive, electrically noisy, or explosive environments.)

Corrosive environments require corresponding corrosion-resistant materials.

14.3 Gripper Actuation

The power sources available for gripper actuation include:

Vacuum	Electromagnets
Air pressure	Electric solenoids
Hydraulic pressure	Springs
Electric motors	Gravity

Each of these has been discussed above, either in general or in an example. Hydraulic pressure is used only in hydraulic cylinders (linear and rotary), but air pressure is used to drive diaphragms, expanding tubes, bags, and other flexible enclosures as well as in cylinders. It is also used for open blasting with air jets.

For friction-gripping sensitive objects it may be important to regulate the pinching force closely despite dimensional tolerances in the part. One way is to engage the jaws with an electric motor, include a strain-gauge force cell or motor current measurement to measure the force, and servo-control the motor until the force is as specified. This is the most expensive and unreliable substitute for an air cylinder with a pressure regulator that I know of. An even more uniform force technique, with almost no hysteresis from friction or otherwise and no sensitivity to power supply fluctuations, is to apply the force from a spring or weight and use any powered device to retract the spring or weight.

14.4 Misalignment

A gripper approaches a workpiece misaligned with it, the gripper moves the workpiece toward a workholder with which the workpiece is misaligned, and each misalignment is in one to six axes. Correcting for this misalignment is a fundamental problem of all gripper designs.

The sources of this misalignment include one or more of the following:

1. Inaccuracy of the robot
2. Nonuniformity of the workpiece
3. Cumulative nonuniformity of a stack or row of workpieces which rest against each other.
4. Inaccuracy of the workholder (tote, chuck, machine, rack, etc.)

There are variations in the initial and final dimensional relationships among source workholder, part, gripper, robot, and receiving workholder. If the workholders hold stacks or rows of parts, the position of each part is different from that of its neighbor and the position is affected by the cumulative tolerance of the parts. In some cases the position of the part in the workholder is accurately known, as a bar in a lathe chuck; in other cases it may vary widely, as a raw casting in a loose nest in a tote box. The tolerances of tote boxes or other carriers are likely to be loose.

There is a transient moment when the part is held by both the workholder and the gripper; so the tolerance on gripper position is zero.

What is done to correct this misalignment so that the robot will work without parts jamming and without damage to robots, parts, or workholders?

14.4.1 Passive self-alignment

The first solution to this problem is to cage a floating gripper with a low seating or centering force to a nominal starting position. Then you let it self-align with the workpiece as a result of interference forces between their surfaces. The ability of a gripper to self-align is referred to as "permissiveness" or "compliance." The permissiveness designed into a gripper may have one to six axes, depending on the detail needs of the problem. A few examples are given:

Figure 14.9 shows a vacuum gripper used to destack and restack disks (actually computer memory disks). Vacuum gripper 1 (three cups) is suspended by gravity from three cones 2, which fit partway into three holes 3 in robot attachment bracket 4. (To be pure MinCD they would be cone in hole, V in slot, and ball on flat.) As the gripper, with or without a carried disk, is lowered against the stack, it is stopped by contact with the stack, the bracket lowers more until limit switch 5 commands it to be stopped, and the gripper is free to self-align with the stack both vertically and radially by centering post 6. The same device can be used horizontally if springs replace gravity. There are six axes of permissiveness.

The gripper not only fits the top disk passively but, under control of limit switch 5, follows a growing or diminishing stack of disks. This is *active* self-alignment and will be discussed below.

Figure 14.6 shows the general principle of inclined surfaces on gripper and workpiece guiding the gripper into position.

Figure 14.10 shows the general principle of pilot pins 1 on the floating gripper 2 engaging corresponding pilot holes 3 on the workholder or its associated machine tool or equivalent.

Aside from grippers I have twice used this action to align delicate electrical connector pins with their sockets. Once was in a Navy intercom which was roughly slammed into its enclosure by a sailor. No sailor in any storm ever bent a pin. Again in a robot automatic conveyor test system in which each passing rack of personal computers paused for robot loading and was tested, this pilot-pin system was successfully used to make multiple connections with the racks. I also used the system in an automatic tool changer for a lathe in which the tool holders had been designed for skilled human alignment and could not be modified to provide a set of substantial entry inclined planes for self-alignment.

A floating gripper can also permit the workpiece to self-align with the workholder. An extreme case is illustrated in Fig. 14.8. The workpiece is an 800-pound, 1-ft-diameter lathe fixture 1 which must

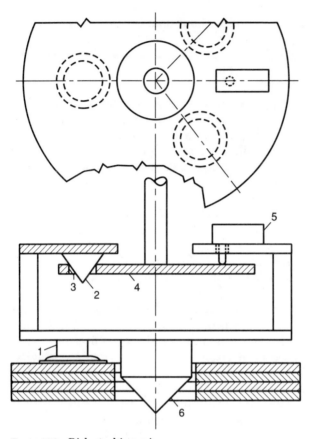

Figure 14.9 Disk-stacking gripper.

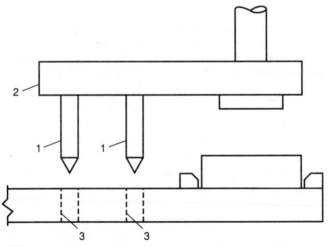

Figure 14.10 Pilot-pin alignment.

enter a diaphragm chuck 2 with a diametral clearance of only $1/32$ in. Both have very short, steep entrance chamfers. The initial position of the fixture centerline 3 is set below the chuck centerline 4, and most of the weight of the fixture is balanced with air cylinder 5 having regulated air pressure. A gentle push sends the fixture into the chuck. Other permissive elements (not shown) permit self-alignment on the other axes.

Another self-aligning action which may be useful in certain cases is the action of shaped gripper clamping jaws in forcing the workpiece to self-align with the jaws instead of the other way around.

If the robot designer is permitted to influence the machine or workholder design, life can be easier for all concerned.

Note that the tolerance to which self-alignment works is plus or minus zero. The workholder and the robot gripper can each clamp the part with zero misalignment stress on anything.

If no provision is made for self-alignment, it either happens anyway—by bending the robot or the workholder or the workpiece or all three—or else there is a jam.

14.4.2 Active self-alignment

Passive self-alignment is limited by the length of the inclined surfaces on the workpieces and workholders. Often the misalignment to be corrected is larger than the self-aligning scope of these surfaces.

Active self-alignment is sensing the part or the workholder and driving the *robot axes* under sensor control until the alignment is close enough for passive self-alignment to complete the job.

The simplest version of active self-alignment is to program the robot to a position definitely short of its final position and then transfer control to limit-switch sensors. Then, simultaneously or sequentially, the robot creeps in the programmed directions until the limit switches tell it to stop. The limit switches may seek contact points either on the workpiece or on the workholder or on its machine. Figure 14.9 illustrates active self-alignment in the vertical direction. I have used this system on precision machine tools where the inaccuracy was almost entirely within the giant robot which loaded them and also on moving wire-frame storage racks whose location was uncertain to inches.

The opposite extreme in complexity is the use of a video camera sensor whose image is analyzed by computer to generate robot error signals.

All kinds of position sensors are useful. They may be electric, pneumatic, or optical. Air nozzle sensors have the benefit of cleaning their targets as they approach (Fig. 14.8). Different kinds of sensors may be used for different axes.

It is easier to do active self-alignment with Cartesian robots than with jointed-arm robots since the sensor axes can correspond to the robot axes one for one.

No matter how one would like to standardize robot grippers, they are a continual drain on engineering budgets. However, the techniques become standardized, and your skill continues to grow; so the cost continues to come down.

Chapter 15

Selecting Power Forms

You have available a great variety of forms of power and devices for transmitting, storing, and using it. A single machine may use two or more forms; consider your car. The choice of optimum form, optimum combination of forms, and optimum parameters (e.g., voltage, pressure) may be far from obvious. Your choices will help determine the value of your machines to your customers. I suggest that you reread Chap. 8 at this point.

This chapter will try to organize this mine of resources to help you make the best choices for your machines.

15.1 Forms of Power

Power is available in many forms. Some may surprise you, but you will soon see examples of all in regular use. You will rule out many for the particular kind of machine on which you now work (explosives and hydraulics are unlikely to be useful in sensitive instruments), but you cannot foretell your future, and learning the subject will deepen your insight into the field of machines.

1. Electricity
2. Compressed air (pneumatics)
3. Pressurized oil (hydraulics)
4. Vacuum
5. Combustion engines
6. Explosives
7. Human muscle

8. Heat
9. Sunlight
10. Wind
11. Gravity
12. Elasticity (springs)
13. Inertia storage
14. Utility water
15. Nuclear

15.1.1 Electricity

Where utility electric power is available, electricity is always the first choice for primary power. Even when the machine is hydraulic or pneumatic, electric motors are the prime movers for the pump or compressor.

Where accurate speed or position control is needed, electric servos are usually used; they are challenged in a few applications by hydraulic servos.

Instrumentation and control are usually electrically powered.

Pneumatic instrumentation still contends for some of this work, particularly where process control valve positions are to be varied; diaphragm valve operators are hard to beat. Within hydraulic and pneumatic systems some logic is performed by hydraulically and pneumatically operated valves; however, hydraulic servos always have electrically operated servovalves.

Fluidic controls, both analog and digital, generally using compressed air, compete with electrical controls in explosive, hot, electrically noisy, and radioactive environments.

In addition to motors and controls, electricity is used to power solenoids and voice coils which move other devices, such as valves. (Voice coils are made in a range of forces from dynes through thousands of pounds.)

Piezoelectric crystals and magnetostriction devices are used to move small distances, usually but not necessarily at high frequency. Ultrasonic power is usually generated by one or the other. The centerless grinder adjustment drive described below uses a large nickel rod in a coil as a magnetostrictive drive.

Linear electric motors are now available to drive linear motions without intermediate gears.

Electricity is used for heating by way of resistance heaters, arc furnaces and welders, plasma arcs, lasers, ultrasonic generators, induction heaters, dielectric heaters, and radiant heaters.

Most electricity is produced by rotating generators. Some is produced chemically by primary batteries for portable devices (flashlights, watches, calculators). Storage batteries store electricity from generators for use during generator outages. Solar cells generate electricity from sunlight but are still an expensive source used only when other sources are unavailable, as in satellites, or as a gimmick such as powering pocket calculators. They require storage batteries for operation during dark time.

15.1.2 Pneumatics

Within a factory, compressed air is a utility which can be piped to any location; so it requires little equipment at each application in order to use it. The essential local equipment is for cleaning, pressure-regulating, and lubricating the air. Filter-regulator-lubricator (FRL) sets are common and inexpensive. Filtering may use anything from a simple porous member to refrigeration desiccators and coalescing filters, depending on the amount of moisture in the air and the sensitivity of the pneumatic devices.

I have been told by factory managers that "compressed-air systems are no darned good; they get dirty and fail. Hydraulics is much more reliable." This puzzled me for a long time, because the opposite is true, but I finally figured it out. Everyone knows that hydraulic systems are extremely sensitive to dirt and therefore must be carefully maintained. Therefore they *are* carefully maintained, and therefore they *are* reliable. Pneumatic systems are relatively tolerant of dirt and moisture, and failure to clean the filter or resupply the lubricator leads only to gradual degradation; so in an undisciplined shop there is neglect and eventual failure.

Because factory air is a utility, most factory systems are at standard 80 to 100 lb/in^2. Instrumentation systems use 15 lb/in^2. Special systems such as in fighter aircraft may use several thousand pounds per square inch to reduce the weight of components.

Conventional uses for pneumatic power include:

- Linear-motion cylinders
- Rotary-motion cylinders
- Positive-displacement motors
- Turbine motors, usually in hand tools
- Liquid spraying
- Air-jet cleaning
- Air hammers (jackhammers, chisels, paint chippers, nail drivers)
- Instrumentation (sensors, controllers, data transmission)

Outside the factory it is common to use engine-compressor sets as pneumatic power supplies for large amounts of power (e.g., outdoor concrete breaking with jackhammers) and electric motor-compressor sets for small amounts of power (e.g., paint spraying). Portable compressed-air tanks and carbon dioxide tanks are used to power hand tools on construction sites to avoid the necessity of long hoses.

Pneumatic power is energy-inefficient because in most devices other than turbines the energy of compressing the air is dissipated when the air is exhausted. Pneumatic power is used where its other advantages outweigh this inefficiency.

15.1.3 Hydraulics

The great virtues of hydraulic power are as follows:

15.1.3.1 Force. Very large forces and torques can be generated in small packages.

15.1.3.2 Power. Power levels from fractional horsepower to thousands of horsepower are possible with commercial components. For some devices, such as automobile jacks, human pumping power is the simplest source.

15.1.3.3 Incompressibility. Oil is approximately incompressible; so a hydraulic system is approximately inelastic and holds its position when the valves are closed. (In high-frequency servos, actual compressibility is a limitation on frequency response.)

Conventional uses for hydraulic power include:

- Linear and rotary cylinders
- Rotary motors
- Isostatic pressing

Unconventional uses for hydraulic power include expanding bearing inner races for easy removal by pressurizing a groove in the shaft inside the race.

I once used hydraulic pressure to squeeze a thin-walled sleeve inward against a sliding shaft to serve as a brake. The device was an extremely-fine-resolution linear drive for adjusting centerless grinders, driven by magnetostriction of a 2-in-diameter nickel bar. All the brakes which had been tried bent the shaft sufficiently to disturb the setting.

An enormous amount of ingenuity has been expended to develop highly sophisticated hydraulic control valves for regulating flow, direction, and logic.

15.1.4 Vacuum

Vacuum power is generated either by pumps or by aspirators. The aspirators are usually powered by compressed air, although water flow is sometimes used.

The conventional uses of vacuum include:

- Cleaning
- Gripping
- Low-power turbines and diaphragm actuators
- Manufacturing environments (furnaces, glove boxes, evaporators, etc.)
- Conveyor tubes

I once made a machine to grow larvae into moths for a pest control program using sterile male insects. The hard part was to transfer the larvae, one at a time, without injury, from a chamber in which they were feeding, en masse, on lettuce leaves to individual cups in which they would continue to grow (individual cups, because they became cannibals). We made the chamber a pressurized glove box and moved a flexible tube by hand from larva to larva. The other end of the tube was mechanically moved from cup to cup. Airflow from inside the chamber to the outside world sucked up each larva and delivered it safely to its new home. The entire outside world was our vacuum chamber. I copied the idea from Jack Rabinow of the National Bureau of Standards, a great inventor, my former boss, and a close friend, who once responded to a challenge to make a 100 lb/in^2 vacuum. He used it to slam steel balls into an assembly.

15.1.5 Combustion engines

Combustion engines include Otto-cycle, diesel-cycle, and gas-turbine engines.

Some combustion engines drive electric generators, and some drive mechanisms directly. Directly driven mechanisms include tiny model-aircraft propellers, portable tools such as garden cultivators and blowers, vehicles of all kinds and sizes from motor scooters to ships, and transportable air compressors and hydraulic pumps.

15.1.6 Explosives

Explosives deliver a quantity of energy quickly; so their instantaneous power level is high. The commercial name is "powder-operated devices." Typical uses are:

- Stud drivers
- Emergency wrenches
- Bolt and cable cutters
- Actuators

Stud drivers are common construction tools. They are powered by blank pistol cartridges.

Bolt and cable cutters are small guillotines with a cartridge-powered piston driving the blade. The cartridge is usually fired electrically. These cutters are widely used in space vehicles to release other devices after launch.

Actuators are pistons and cylinders or bellows which are operated by the combustion gas of the powder charge. The powder may be relatively slow-burning.

Explosives are also used in industry for laminating metals and for forming metals.

15.1.7 Human muscle

Although we think of machines as powered, many of them are driven by human muscle for effectiveness, economy, emergency, or sport. Examples are:

- Hand tools for shop, office, and surgery
- Spring winding and weight lifting (bows and crossbows, mechanical clocks, exercise machines, hoists)
- Pedaled bicycles and boats and human-powered airplanes
- Oar-driven boats (recreation, racing, and lifeboats)
- Small pumps (water, air, and hydraulic)
- Emergency radio generators

Animal power is not included here because I know of no remaining opportunities for using it in machine design. At one time it was the major source of useful power.

15.1.8 Heat

Heat is used for chemical and metallurgical processes, space heating, and welding. Most heat is obtained from electricity and from combustion. Solar furnaces using sunlight focused by large reflectors have been employed experimentally.

Electric heat is described above. Combustion heat uses fuel and oxidizer. The most common oxidizer is air, but sometimes pure oxygen is used, and for special purposes fluorine or other chemicals may be used. Most fuels are liquid or gaseous, but aluminum is used as a fuel in thermite welding. Most gas fuel for fixed installations is provided by public utilities; gas fuel for portable equipment and remote installations is compressed into tanks.

15.1.9 Sunlight

Sunlight is converted to electricity by solar cells and stored in storage batteries for dark periods. Such power is now used for low-power installations remote from utility power. Low-power instruments transmitting data by radio are an example. Almost all space vehicles depend on solar cells. Such photovoltaic use of solar power is a major hope for a pollution-free and inexhaustible source of major quantities of power. The present limitation is the cost of the solar cells.

Using sunlight for illumination is more in the realm of architects than that of mechanism designers.

You may be able to propose machines to be solar-powered which formerly could not be made.

15.1.10 Wind

Wind-driven generators of less than 1 kW have long been used as sources of electric power in farms not served by utilities. They are usually backed up by storage batteries for low-wind and peak-load conditions.

Arab oil profiteering was the incentive to develop utility-scale wind generators, but they have never been economically successful; most were paid for by tax incentives originated to promote R&D. The windmill industry prays for increased oil prices. I met the president of one such company which raised millions of dollars of capital based on *predictions* of *future* oil price increases (which did not materialize, I am glad to say).

Aerial bombs use tiny wind-driven propellers to operate and power safety and arming devices.

Wind-driven sailboats in both cruise ship and racing yacht sizes are under continual improvement despite having started at least 3000 years ago. New materials such as aluminum, Dacron,* and fiberglass laminate have permitted many of the improvements, and aerodynamic knowledge from the aircraft industry has contributed greatly.

Gliders are not exactly wind-driven but are lifted by thermally generated or mountain-deflected vertical winds and then driven by gravity. Balloons are wind-driven.

Wind power derives from earth heating by sunlight.

15.1.11 Gravity

Gravity provides utility-scale power via falling water. (The water was first raised by evaporation due to heating by sunlight.) It is used in utility-scale energy storage by pumping water from low rivers to high lakes during off-peak-load hours and letting it flow back down during peak-load hours.

Most other uses are obsolete. Water clocks and hourglasses used gravity. Some catapults were gravity-powered. Tower clocks, grandfather clocks, and some laboratory instruments used weights raised by human power and pulled down by gravity to drive them. The only current use I know of is to provide a reliable water supply from an elevated tank. One thing about gravity: it never fails. Please see the story of the heart-lung machine in the subsection "Utility Water" below.

15.1.12 Elasticity

In ancient days some military catapults were powered by torsion springs; bows and crossbows driven by twin cantilever springs are still being made. Before electric clocks, everyone had an alarm clock and a wristwatch powered by spiral springs wound by hand. "Self-winding" wristwatches wind their springs by the swing of pendulums inside the watches, the pendulums being swung by gravity and random arm motions. Today there are still windup toys, kitchen and other timers, and a few instruments. Springs are used to store energy in starters for small gasoline engines. Rubber bands in torsion still power model airplanes. Transient energy storage by metal or rubber springs is standard in every vehicle suspension. Springs using compressed gas instead of stressed solid have long been used in vehicle suspensions.

---

*Dacron is a trademark of E. I. du Pont de Nemours & Co.

Elastic energy storage remains a living technique for you to consider in designing your machines.

15.1.13 Inertia

Flywheels can store substantial amounts of energy. The usual use for flywheels is for transient storage to smooth the action of reciprocating engines and to supply peak loads in a machine cycle so that the machine's motor or engine can be sized for its average load.

There are longer-storage-time uses:

1. In friction butt welding of shafts, the energy for each weld is loaded into flywheels coupled to the shafts and then is transmitted to the weld as the shafts are pressed together.

2. Much work has been done in Switzerland to use flywheels as regenerative energy stores for buses in hilly country.

3. Airplane reciprocating-engine starters were once made in which a hand crank loaded energy into a flywheel, which then dumped it into the engine.

Research has been done to make flywheels competitive with storage batteries as energy stores. The limitation to the energy capacity of a flywheel is its tensile strength in resisting centrifugal force. Work has been done to apply modern high-strength materials.

When I was in the proposal end of the space business, my company put me in charge of the proposals for Project Prospector, which was to land 2000-lb unmanned packages on the moon. I proposed a moon surface vehicle in the form of a sphere whose primary power was solar cells and whose energy store was a large flywheel. We built a model which worked great. Not only did that wheel store energy but it necessarily stored angular momentum as well, and when a brake was applied to it, that sphere really *turned* regardless of any obstacle. (It tried to climb a vertical wall, but it lacked traction.) There was great enthusiasm, my chief engineer ordered absolute secrecy, and I had the bittersweet experience of having to refuse to talk to a reporter from *Aviation Week*. Then the Russians flew Gagarin, manned space became the order of the day, and the project was killed. But that flywheel really worked!

15.1.14 Utility water

Tap water is a little used source of hydraulic power. The only devices using it with which I am familiar are a lawn-sprinkler crawler and a back scrubber for the shower in which a water motor turns brushes and moves them up and down (I have one; it works very well).

It was partly for its reliability that I used gravity-backed water power to drive the heart-lung machine mentioned elsewhere in this book. I powered the machine with a positive-displacement water engine using city water as the energy source. (There was a storage tank on a hill nearby.) One liter of water displaced one liter of blood. Measuring water flow measured blood flow. The machine was quiet and explosionproof. It worked fine, but conventional electric motors ultimately took over.

I can conceive of kitchen appliances such as an orange squeezer driven by tap-water piston and cylinder, and I suspect that a garage car lift might cost a little less. The cost of plumbing installation is an obstacle, of course. Consider this neglected resource an opportunity to invent.

15.1.15 Nuclear

Aside from large utility and ship propulsion reactors, the Russians have launched at least one space vehicle powered by a small reactor, and we have at least proposed small, long-lived power supplies in which the heat from decaying radioactive material generates electricity in a thermopile.

One purpose of this recitation is to encourage you to include in your thinking some of these unconventional power sources so that your machines may be more capable or less costly than they might otherwise be.

Chapter 16

Backlash

The looseness between two meshed gears is called backlash. Similarly, almost every motion-transmitting chain of two or more moving parts has some looseness (i.e., backlash) in the junction between each pair of parts. When a force is applied to one end of the chain, it causes all the loosenesses to shift in one direction so that the parts can transmit the force. When the force reverses, the loosenesses shift in the other direction. The same effects occur with torques as with forces. The shifting looseness is called backlash.

In power trains which transmit power in one direction only, backlash does no harm since the looseness never shifts. An example is the hand drives for watches and clocks.

In reversible positioning drives the shifting looseness—backlash—causes shock, noise, wear, inaccuracy, and servo instability. It is important to be able to take it out.

Backlash is eliminated by introducing a steady force or torque, such as a spring, gravity, or power load, into the chain to move the total backlash into one direction and hold it there. *This antibacklash force must be large enough to hold the backlash at the same end even when the applied force or torque opposes it.* The antibacklash force is a seating force.

A cost of using an antibacklash seating force is that the capacity of the gears, bearings, and other links in the chain must double their capacity to do their jobs without the antibacklash seating force. In one direction they must carry both working force and seating force, and the seating force must be at least equal to the working force in order not to be overcome by the working force. In the other direction the seating force does the driving.

All this is pretty abstract; so let us turn to examples.

1. Figure 3.24 shows spring 9 generating the antibacklash force and nut 8 providing a traveling base for the spring to react against. The string of loosenesses is:

- Lead screw to nut 1
- Rod ends 3 from nut 1 to gimbal 2
- Gimbal 2 to table extension 4
- Spring 9 to nut 8
- Nut 8 to lead screw

2. Figure 16.1 shows preloaded angular-contact ball bearings. Figure 16.1*b* shows the bearings in low-pressure contact with a small gap 1 between the outer races. Figure 16.1*a* shows the bearings assembled

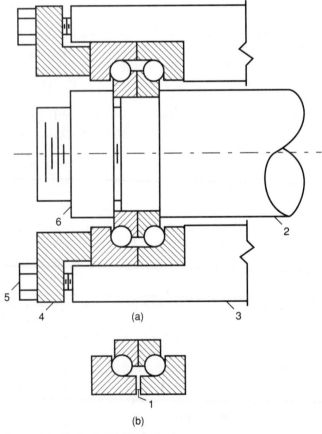

Figure 16.1   Preloaded ball bearings.

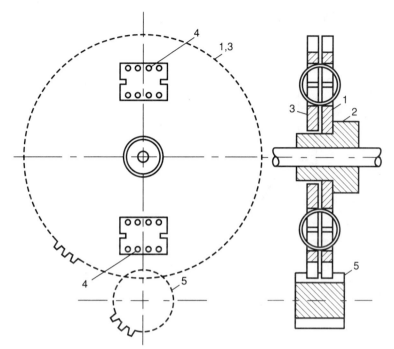

Figure 16.2 Antibacklash split gear.

on shaft 2 in housing 3 and with the outer races forced together by ring 4 and screws 5. The antibacklash force is the elastic compression and indentation of the balls and races. This compression is very small; so the bearings are made with great accuracy and are provided in matched sets in order to control this small displacement. Nut 6 eliminates any axial looseness between the shaft and the inner races.

3. Figure 16.2 shows a commercial split antibacklash instrument gear meshed with a simple pinion. Gear 1 is fixed to hub 2. Identical antibacklash gear 3 floats on hub 2 and is coupled to gear 1 by antibacklash springs 4. Both gears mesh with pinion 5. Springs 4 force gear 3 against one side of the mating pinion tooth and force gear 1 against the other side of the same tooth.

4. Figure 16.3 shows antibacklash bevel gears. Ball bearing 1 isolates spring 2 from the pinion rotation. No additional antibacklash gear is required. I have seen a large set of bevel gears in which a hydraulic cylinder, instead of a spring, provides the antibacklash force.

5. Figure 16.4 shows an antibacklash rack-and-pinion drive. Shaft 1 turns pinion 2, which meshes with rack 3. Torsion spring 4 is coupled to shaft 1 at one end and to gear 5 at its other end. Gear 5 is sup-

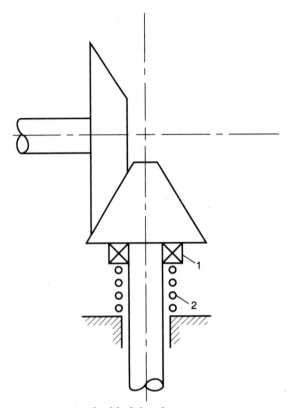

Figure 16.3 Antibacklash bevel gears.

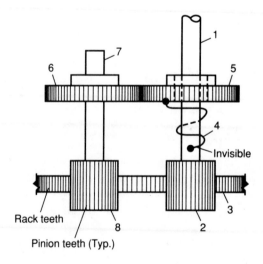

Figure 16.4 Antibacklash rack-and-pinion drive.

ported on bearings by shaft 1 but is angularly free from it. The torque on gear 5 is transmitted to gear 6, shaft 7, and pinion 8. The spring causes equal and opposite forces between pinion 2 and the rack and pinion 8 and the rack. The spring could be higher up the drive gear train, closer to the motor, if backlash elimination between the motor and pinion 2 were also wanted. The cost would be a duplicate gear train from the spring down to pinion 8.

6. Figures 16.5 and 16.6 show a rack-and-pinion system with antibacklash and with other features which may be of value to you some day. Rack 1 is coupled to antibacklash pinions 2, 3 via large-

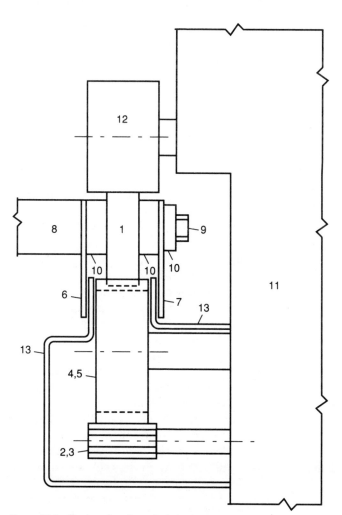

Figure 16.5 Enclosed rack-and-pinion system.

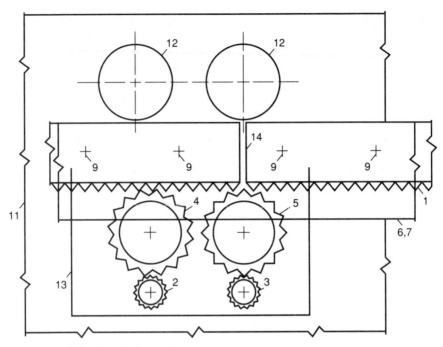

Figure 16.6 Side view of Fig. 16.5.

diameter idler gears 4, 5, whose backlash is also taken up by the antibacklash spring between the pinions. The purpose of the idlers is to permit dirt guard strips 6, 7 to hang well below the rack teeth.

The rack and guard strips are mounted on fixed structure 8 by bolts 9 and spacers 10. The guard strips can be flexible, urged together by springs or inflated air tubes, and separated locally by the passage of the end of gear enclosure 13. Such guard strips and gear enclosure offer protection of the rack-and-pinion teeth from the environment. Protection of the upper track edge of the rack from dirt which would lift the wheels and jam the gear teeth is easily obtained by brushes on carriage 11, perhaps supplemented by air jets.

The pinions, idler gears, and the pinion drive are mounted in traveling carriage 11. Carriage 11 rolls on wheels 12 on the smooth edge of the rack which serves as a track. This configuration is a MinCD. It assures uniform meshing of pinion and rack regardless of rack straightness variation. If a separate track not integral with the rack were used, there would be nonuniform meshing of pinion and rack due to nonparallelism and spacing variation between the rack and the separate track.

Because racks are made in relatively short lengths and traveling machines such as robots and cranes may travel hundreds of feet, provision must be made for rack joints. Racks are usually made such that if the end teeth are properly spaced, a small gap 14 appears (Fig. 16.6). If redundant wheels 12 are provided, then one wheel will carry the load when the other is crossing the gap. This is not done on railroads, with the result that there is a shock when each wheel crosses a joint. (The railroad companies sometimes weld a mile or more of track to minimize the number of such shocks.) These redundant wheels are another example of valuable RedCD.

Compare Fig. 16.6 with Fig. 2.15. Figure 2.15 does not show redundant track followers in order to keep the sketch simple, but in practice they are used.

After studying these examples, you should reread the first four paragraphs to understand, as general principles, what you now understand as hardware examples.

Chapter 17

Hype

We are exposed to an unending stream of slogans, exaggerations, and downright lies. They are composed and propagated by people for their own self-interest, to persuade rather than inform, in order to:

1. Persuade others to buy their products or services
2. Acquire more power via publicity and the appearance of wisdom
3. Persuade others to work harder or better
4. Persuade themselves of their own importance (ego trips)
5. Promote a political program

It is useful to apply our knowledge and judgment to slogans so that we can resist confusion and unreasonable persuasion.

I have made a collection of slogans of this sort which have been applied to engineering and list them below. You may want to add to the list—I should be glad to hear from you.

- Postindustrial society
- The Information Age
- Zero defects
- Factory of the future
- Man in space
- Robotics (as if it were a fundamentally new technology)
- Artificial intelligence
- Group technology
- Value engineering
- Know-how

- Total-quality management
- Consumer society
- Just-in-time
- Welfare society
- Megacompany
- Revolution in . . . (technology, management, etc.)
- *Today's* engineer, management, faster pace, etc. (implying that something has changed which has not)
- Any new word for an old thing

The list does not include presidential campaign slogans, commercial product slogans, theater advertisements, and other hype directed at everyone, not just engineers.

I was tempted to give my own analysis of the items on the above list, but since you are trained engineers too, I'll leave the analysis to you.

Another class of deception which we should learn to recognize and correct in our own minds is journalistic distortion which operates by:

- Selecting or ignoring
- Emphasis or deemphasis (e.g., by position in a newspaper, by amount of allocated space or time on radio and TV)
- Exaggeration or minimization
- Opinion disguised as fact
- Praising or blaming words (e.g., "freedom fighter" versus "terrorist")
- Reached-for superlatives

The power of journalism is used to propagate the opinions of the journalists, promoting hidden agendas.

Journalism (TV, radio, and print) is paid for by the advertising business to attract our attention to the advertising; so "news" is selected and written for drama rather than for importance.

Chapter 18

Product Deterioration

As soon as your product comes into existence, it starts to degrade. Some deterioration starts even during manufacture. The product becomes shabby and damaged, and its performance deteriorates.

What are the processes and events which cause deterioration, and what can your design do to prevent, minimize, and correct it?

18.1 Spontaneous Deterioration

Some deterioration occurs even when the product is not even being used, whence the expression "shelf life." Most of this is the result of slow chemical processes within the product and between the product and the atmosphere. Among these processes are:

- Decay of adhesive bonds
- Oxidation of surfaces
- Reaction of surfaces with humidity, ozone, and pollutants
- Chemical reactions between adjacent materials in the product (e.g., plastics, lubricants, fuels, chemicals)
- Electrolytic corrosion due to contact between dissimilar metals

Creep of metals and plastics under built-in stress within a part, assembly stress among parts, and cyclic thermal stress result in dimension changes and loosening of fasteners.

18.2 Attacks during Shipment

While the product is being shipped, it may be subject to attacks different from those in storage or in use. Among these are:

- Shocks from impacts and dropping
- Vibration of transporting vehicles
- Tipping and turning over
- Concentrated lifting and pushing forces
- Weather (snow, rain, hail, cold, heat, sunlight, salt spray)
- Pollutants, from fingerprints to acid rain
- Low atmospheric pressure in airplanes
- Dust and dirt (In medical products "dirt" includes germs.)
- Fungus
- Sabotage by vandals and by enemies of all sorts
- Heat

18.3 Environmental Attacks

Some of the attacks during shipment continue after the product has been received, unpacked, and put into service. "Unpacking" includes the removal of protective enclosures and coatings used to protect the product during shipment and so exposes the product to such attacks. Furthermore, it may be necessary to replace shipment protective coatings with lubricants which also act as protective coatings. Delay results in a transient exposure of parts without protection for an uncertain time.

Additional environmental attacks which may occur in service include:

- Wind and dust erosion
- Lightning
- Small electrostatic discharges and other electrical noise
- Electrolytic corrosion from stray electric currents
- Electrical surface leakage from surface accretions and electrical leakage through solids from moisture absorption
- Changes in light reflection from surface accretions
- Prolonged sunlight
- Corrosion and erosion by fresh, salt, or polluted water (Corrosion is not just of metals; even stone corrodes.)
- Corrosion by polluted air
- Corrosion by solid, liquid, and gaseous chemicals, including per-

spiration, food, drink, blood (particularly in medical products), wastes
- Mechanical interferences from material accretions (e.g., dust, coffee spills)

18.4 Wear

Both normal and abnormal use removes surface material. The effects are to change surface texture, remove material, and remove protective finishes. Engineering inattention to wear and corrosion causes most product demise.

18.5 Abuse

Superficial abuse degrades appearance but does not interfere with function. It includes scratches and dents and dirt.

Serious abuse includes failure to perform preventive and corrective maintenance, use of improper materials (e.g., fuels, lubricants, chemicals, replacement parts), and damage which interferes with function or accelerates environmental damage. Such damage may result from impact or from overload.

18.6 Design and Manufacturing Errors

Any new product may have errors which are exposed by service faults and which require changes in the field or recall to the factory.

18.7 Modification by User

The product's user may add or remove or modify parts in ways which amount to second-guessing your design. For example, punch presses have been modified and then caused injuries which resulted in product liability suits against the original manufacturer.

18.8 What Can You Do about It?

1. Visualize shipping and service conditions as you design and try to forestall deterioration by design.

2. Show your design to chemists and electrical engineers in your company, and ask them to anticipate deterioration modes. Your material and component vendors' sales engineers make valuable free consultants.

3. Make it easy to do preventive and corrective maintenance. Make

maintenance spaces easily accessible; this will reduce the temptation not to bother.

4. Provide indicators which predict trouble (thermometers, voltmeters, level gauges, etc.). Provide warning signs and instruction signs. Many standard signs and coded tapes can be purchased. (3M)

5. Prepare service and instruction manuals which are easy to understand. Include symptoms which predict trouble. The manuals are part of the product.

6. Include packaging and the shipping mode as part of the product design.

7. Offer replacement materials and parts as your own products, or at least carefully specify them in the manuals.

Chapter 19

Electrical and Mechanical Technologies: Competition and Cooperation

19.1 History

The competitive progress of the two technologies has been a lopsided win for the electrical. However, there are functions which are inherently the turf of mechanical design, and machines are increasingly combinations of electrical and mechanical technologies.

Mechanical design has been going on since the first stone ax. It really took off with mechanical clocks in the seventeenth century and has been making progress ever since, but with few dramatic breakthroughs.

Electrical technology didn't exist at all except as a laboratory curiosity until the nineteenth century. The first commercial use for electricity was the telegraph, which appeared just before the Civil War. Useful motors and generators date from the late nineteenth century. The vacuum tube was invented early in the twentieth century, the transistor a few decades later, and the integrated circuit shortly thereafter. Larger- and larger-scale integrated circuits continue to appear, and these are only the most conspicuous tips of the enormous icebergs of the electrical art.

Success breeds success. The success of the electrical art has led both government and business to invest more and more money in it, thus accelerating its further success. Another result of the electrical success is that the field is competitively more attractive to the best students, who, in turn, accelerate its achievements. Success and failure are both exponential.

19.2 Electrical Takeovers

Direct takeovers of mechanical functions by electrical or electronic devices include the following.

19.2.1.1 Motors. Electric motors operating from utility power replaced steam engines as power sources in factories. Multiple electric motors first replaced line shafting to drive individual machines and then replaced mechanical power transmissions to drive different portions of the same machine.

19.2.1.2 Variable-speed motors. Variable-speed electric motors, electronically controlled, have replaced gearboxes, cone pulleys, and other variable-ratio mechanical transmissions in many classes of machine.

19.2.1.3 Instrumentation and control. Electrical instrumentation and control have almost completely replaced mechanical instrumentation and control. Centrifugal governors, steam engine indicators, and pneumatic process controls are either gone or on their way out. Electronic sensors for almost every physical quantity are commonplace, together with remote indicators and closed-loop automatic controls. Even the engines in our cars have become electronically controlled. Fiber optics is not electrical in nature but is within the electrical field because the input and output devices are electrical.

19.2.1.4 Programming. Electrical programming of machine functions (temperature cycles, sequences, machine tool paths, etc.) has displaced mechanical programming by cams, templates, clockwork, and pneumatically or mechanically sensed punched tapes as in Jacquard looms and player pianos. However, I can tell you of one exception:

When Rabinow Engineering Co. developed the standard U.S. Post Office letter sorter in the 1950s, it was decided that the sorting memory which escorted each letter through the machine must be all-mechanical; it was felt (in 1954) that mechanism was more reliable than electronics and that the Post Office could not undertake electronic maintenance. To this day there are millions of little nylon wheels running through those machines, with extremely high reliability, dropping letters into the proper sorting pockets. The story of that development is given in Ref. 21 in Chap. 20.

19.2.1.5 Computing. Electrical computing has replaced mechanical computing. It has taken over desk calculators, military fire controls, industrial-process controls, and domestic-appliance controls.

Two mechanism technologies have been made totally obsolete by electrical computing: mechanical digital and mechanical analog computing. The digital technology included the desk calculator and punched-card machinery. The analog technology included ball-and-disk integrators, differential-gear adders and subtractors, 3D cam memories, and trigonometric linkages. The planimeter remains. Some pneumatic analog computing for process control survives because it is inherently explosionproof and because pneumatic actuation of modulating valves is hard to beat. Pneumatic digital control (fluidics) survives for severe environments (explosive, heat, radiation). All these are special cases; electronics has won.

Paradoxically, electronic computing has provided the greatest single advance in mechanical engineering: computer analysis of mechanisms and structures (including finite element analysis, FEA), computer-aided drafting, and computer-aided manufacturing.

It is an historical oddity that the first effort to make a large-scale programmed digital computer was the mechanical "analytic engine" invented by Charles Babbage, a mathematics professor in Cambridge University. He received a grant from Parliament and tried to build it between 1820 and 1840, but he could not get the parts made to sufficient accuracy with the manufacturing technology of the period.

19.3 Mechanical Instruments

The principal measuring device which seems to be permanently mechanical is the clock. Every clock is a *mechanical* oscillator, a feedback amplifier to replace dissipated energy, and a counter. (An escapement is a feedback amplifier.)

Quartz clocks and watches use the *mechanical* oscillation of a quartz crystal and *electrical* feedback power and counting. All-electric oscillators have more energy losses (lower Q) and therefore less sharpness of tuning than mechanical oscillators such as a balance-wheel–spring combination (a torsion pendulum) or a pendulum-gravity combination.

The two ultimate clocks are earth rotation, which can defensibly be called a mechanical phenomenon, and the atomic clocks, or masers, which use the mechanical oscillation of atoms in their molecules but provide little employment for mechanism designers.

There are other mechanical instruments which seem to have a long life ahead of them. The gyroscope is one, although laser gyros have appeared. Mechanical gimbals support steerable optical telescopes for surveying and astronomy, radio telescopes, and radar antennas (other than phased arrays).

19.4 Uses for Electricity in Machines

Electricity is used for four principal functions in machines:

19.4.1.1 Transmission of mechanical power. Electric generators convert primary mechanical power sources, mostly water, steam, and internal combustion, to electric power for transmission to user sites. The transmission distance may be less than 100 ft in a ship or a car or more than 1000 mi in a continental power network.

Reconversion to mechanical power is by fixed- and variable-speed motors and their electronic amplifiers and controls, servomotors (variable-speed motors whose speed and position can be controlled quickly and accurately), linear versions of all these motors, and solenoids.

Auxiliary devices involved are switches, fuses, circuit breakers, transformers, fixed and flexible cables, and enclosures.

19.4.1.2 Heating. Electrical heating devices include resistance heaters, welders (spot, arc, plasma arc, laser), radiant heaters, induction heaters, and arc furnaces.

19.4.1.3 Information. Information gathering, storing, processing (computing), displaying, and commanding are where much of the revolution has been. The devices involved are sensors, computers, displays, and the electronic amplifiers which power many of the motors of a machine. These information functions *use* very little power, but they can *control* all the power of any machine. Examples are NC (numerical control of machine tools, replacing tracers and cams), PLC (programmable logic controllers) for control of operating sequence in accordance with rules, control of heating and chemical processes, and built-in microcomputer control of machines such as photocopiers, videocassette recorders (VCRs), fax systems, automobile engines, etc.

19.4.1.4 Miscellaneous. Other electrical functions include electrical-discharge machining (EDM), operation of clutches, brakes, and fluid power valves, vibratory feeders, electroplating, and, of course, lighting.

19.5 The Future of Mechanism

Does all this mean that mechanism is obsolete and that the electrical folks are going to take over all our jobs in the long run?

No way. But it does mean that mechanical engineers must understand the power of electrical engineering and the power of mechanical engineering. They should concentrate on what they do best, *in fact what they do alone.*

We must learn to do more *system* thinking, that is, thinking about the entire electro-mechanical machine and the interrelationships of all its parts and not just about the portion which is purely mechanical.

We should also be aware that the cost of electronic controls, particularly of small built-in computers, will continue to come down and that their software and capability will continue to improve. In these facts the good news for mechanism designers is that this electronic progress will make it economical to build mechanisms which have not been cost-justified before.

The electric power and electronics industries themselves have created needs for mechanisms for which there are no electronic substitutes: utility-scale steam and water turbines and electrical generator construction (e.g., bearings); water, fossil fuel, and nuclear power plant machinery; circuit breakers and contactors of many sizes and ratings; electrically propelled vehicles; drives and changers for computer and audio disks and heads; tape drives and threaders for VCR, audio, and computer tapes; printers, photocopiers, and fax systems; antenna and camera structures and drives; electro-magnetic wristwatch hand drives; a variety of transducers; and a host of manufacturing machines.

19.6 Fields of Mechanism

What fields are reserved to mechanism in the future by the very nature of mechanism?

19.6.1 Engines and turbines

Although most industrial power is first converted to electricity, transmitted electrically, and converted back to mechanical form by electric motors, almost all electrical power first enters a generator as mechanical shaft rotation driven by a heat engine or a water turbine.

Direct conversion of heat to electricity by the thermo-electric effect has shown no promise for large-scale power. Direct conversion of sunlight to electric power (the photovoltaic effect) has made space satellites possible and may some day be cost-competitive in desert areas, but that day seems far off because of the high cost of the devices.

There seems to be no satisfactory primary or storage battery or fuel cell on the horizon for electric power for most cars and trucks. Here the traditional internal-combustion engine remains king. However, electrical control of the engines and transmissions is increasing. Electrical transmissions as alternatives to manual or automatic gearboxes were early contenders but have failed competitively except in such large vehicles as diesel-electric locomotives and some ships.

19.6.2 Moving matter

I was once visited, in my robot company, by the president of a competing company. He had reached his job by way of a Ph.D. in computer science. We made large Cartesian robots which were very impressive to see.

After the $2 tour we sat at my desk and he said, "Larry, these are all very nice, but you can do it all with a chip nowadays. So, really, who needs them?"

When I realized that he actually meant what he said, I laid a paper clip and a pencil side by side on my desk and said, "Let this paper clip represent your dream chip; unlimited memory, infinitely fast, and absolutely reliable. Let this pencil be just a pencil.

"The chip can tell us how high to raise the pencil, when to raise it, and any other command you can imagine about raising the pencil. But the chip won't raise the pencil a thousandth of an inch. It requires a mechanism, which obeys the chip, to actually move the pencil. That's what we make."

He thought for a while and then said, slowly, "You know, Larry, I think I see what you're driving at."

"Moving matter" is pretty abstract; so let's consider some examples:

19.6.2.1 Vehicles.
Vehicles of all sorts are machines to move matter: railroad trains, airplanes, cars, trucks, ships, and bicycles.

19.6.2.2 Material handling.
In factories, warehouses, mines, and construction sites matter is moved with conveyors, cranes, fork trucks, automatic guided vehicles, robots, pumps, and automatic storage and retrieval systems; and, still, by people. The materials so handled include coal, oil, cars, and computer chips.

19.6.2.3 Manufacturing.
Manufacturing anything consists of moving matter, from unloading trucks, to cutting and forming materials, to applying paint, to mixing chemicals, to depositing thin films, to assembling parts. Almost all manufacturing tools and machines are mechanisms. Inspection, control, and some processes such as electron-beam welding, vapor deposition, and electro-plating are electrical. Heat treatment stands alone.

19.6.2.4 Domestic appliances. All are electrically powered and controlled, but an electric toaster must still handle a material slice of bread.

19.6.2.5 Military devices. In the military, matter is moved in many ways at high speeds with hostile intent.

19.6.2.6 Surgical devices. Surgeons have a continuing need for ingenious instruments and machines to examine, displace, and reassemble human tissue. In many cases the *need* arises from the *opportunity* to exploit new engineering and new medical knowledge and technology. Artificial organs and prosthetics, both external and implanted, are examples. Usually the device is electrically powered and controlled, again illustrating the growing prevalence of combination electrical and mechanical machines.

19.6.3 Structures

It is difficult to anticipate electronic water tanks, buildings, or bridges.

19.7 Conclusion

The combining of electrical devices and mechanisms into electromechanical machines has yielded new and improved machines of great economic value and will continue to do so. One result is that the need for mechanism designers remains high. It behooves you, however, to learn enough about the electrical art so that you can communicate with electrical engineers and understand, conceive, and design combined-technology machines.

Chapter 20

References and Bibliography

20.1 Books on Design

1. Whitehead, T. N.: *The Design and Use of Instruments and Accurate Mechanisms*, Dover, New York, 1954. Out of print. Originally published in 1934. This book contains the theory of MinCD and semi-MinCD as well as a general theory of mechanical instrumentation. I was introduced to MinCD by this book.
2. Glegg, Gordon L.: *The Science of Design*, Cambridge, London and New York, 1973.
3. ———: *The Selection of Design*, Cambridge, London and New York, 1972.
4. ———: *The Design of Design*, Cambridge, London and New York, 1969.
5. ———: *The Development of Design*, Cambridge, London and New York, 1981. These short books by Glegg were written by a highly experienced design engineer and contain many insights and lessons in design. They are easy and sometimes entertaining to read.
6. Boothroyd, G., and P. Dewhurst: *Design for Assembly*, Machine Design, Penton/IPC Inc., Cleveland, 1984. Professor Boothroyd is an authority on part feeding and automatic assembly.
7. Hindhede, Uffe, et al.: *Machine Design Fundamentals: A Practical Approach*, Wiley, New York, 1983.
8. Leyer, Albert: *Machine Design*, Blackie, Glasgow, 1973.
9. Greenwood, Douglas: *Engineering Data for Product Design*, McGraw-Hill, New York, 1961. Out of print.
10. ———: *Mechanical Details for Product Design*, McGraw-Hill, New York, 1964. Out of print.
11. ———: *Product Engineering Design Manual*, McGraw-Hill, New York, 1959. Out of print.
12. Chironis, Nicholas P.: *Mechanisms, Linkages, and Mechanical Controls*, McGraw-Hill, New York, 1965.
13. Shigley, Joseph E.: *Mechanical Engineering Design*, 2d ed., McGraw-Hill, New York, 1972.
14. Erdman, Arthur G., and George N. Sandor: *Mechanism Design: Analysis and Synthesis*, vol. 1, Prentice-Hall, Englewood Cliffs, N.J., 1984.
15. ———and———: *Advanced Mechanism Design: Analysis and Synthesis*, vol. 2, Prentice-Hall, Englewood Cliffs, N.J., 1984.
16. Haugen, Edward B.: *Probabilistic Mechanical Design*, Wiley-Interscience, NewYork, 1980.

17. Kennedy, John B., and Adam Neville: *Basic Statistical Methods for Engineers and Scientists*, 3d ed., Harper & Row, New York, 1986.
18. Lipson, Charles, and N. J. Sheth: *Statistical Design and Analysis of Engineering Experiments*, McGraw-Hill, New York, 1972.
19. Carnahan, Brice, H. A. Luther, and James D. Wilkes: *Applied Numerical Methods*, Wiley, New York, 1969.
20. Arora, J. S.: *Introduction to Optimum Design*, McGraw-Hill, New York, 1989.
21. Kamm, Lawrence J.: *Successful Engineering*, McGraw-Hill, New York, 1988. This book contains the first recent presentation of MinCD of which I am aware. It is primarily a career guide and has many essays on design and on the life of a professional engineer. Modesty forbids my saying how valuable it will be to you.

20.2 Books on Information Sources

22. Schenk, Margaret T., and James K. Webster: *What Every Engineer Should Know about Engineering Information Resources*, Marcel Dekker, New York, 1984.
23. Gates, Jean Key: *Guide to the Use of Libraries and Information Sources*, 6th ed., McGraw-Hill, New York, 1988.

20.3 Journal Articles on Design

24. Darwin, Horace: "Scientific Instruments; Their Design and Use," *Aeronaut. J.*, vol. XVII, no. 65, pp. 170–190, July 1913. This is the earliest paper on MinCD of which I am aware. It is now more of a historical document than a source which should be studied for its content.
25. Ferguson, Eugene S., "The Mind's Eye: Nonverbal Thought in Technology," *Science*, vol. 197, no. 4306, Aug. 26, 1977. An excellent study of innovative thinking in engineering, with an extensive bibliography.

20.4 Handbooks

Many handbooks are still published with mathematical tables which have been made obsolescent by the pocket calculator. The calculator generates the desired value and enters it into the calculation when it is needed.

Many handbooks are mostly miniencyclopedias and give a general introduction to their subjects without teaching enough to enable you to do design.

Some handbooks are concentrated design texts and can help you to actually perform design *if* you already know enough to follow the handbook without misinterpreting its formulas.

Other handbooks are sources of huge amounts of numerical data ready to be used. I have found the *Handbook of Physics and Chemistry* to have every physical constant I ever wanted and *Machinery's Handbook* to have every size and dimension of standard mechanical elements I ever wanted.

20.4.1 Handbooks on mechanism design

26. Shigley, Joseph E., and Charles R. Mischke: *Standard Handbook of Machine Design*, McGraw-Hill, New York, 1986. This is a comprehensive handbook containing encyclopedia information, design formulas, and data and including examples.
27. Rothbart, Harold A. (ed.): *Mechanical Design and Systems Handbook*, 2d ed.,

McGraw-Hill, New York, 1985. This is a comprehensive handbook containing encyclopedia information, design formulas, and data. It uses advanced mathematics.
28. Warren C. Young: Roark's *Formulas for Stress and Strain*, 6 ed., McGraw-Hill, New York, 1989. This is an all-encompassing handbook of design formulas with their background theory, but you had better have a course in mechanics of materials behind you.
29. Oberg, E., F. D. Jones, and H. L. Horton: *Machinery's Handbook*, 23d ed., Industrial Press, New York, 1988. This is an extremely thorough handbook of data for mechanical design.
30. *Handbook of Chemistry and Physics*, CRC Press, Boca Raton, Fla., annually. Mostly data of pure science, but some of it is useful in mechanical design.
31. *Materials Selector 1990*, Reinhold, Stamford, Conn. This book is provided annually with a subscription to *Materials Engineering* magazine but can be purchased separately. A detailed and well-organized data handbook on engineering materials, it also contains a number of comparison tables of great value in choosing materials. I think it is one of the most valuable handbooks you can have.
32. *Metals Handbook*, ASM International, Metals Park, Ohio 44073.
33. *AISC Manual of Steel Design*, American Institute of Steel Construction (AISC), Chicago. This manual contains dimensions and data on all standard rolled-steel shapes and much other information.
34. Powell, Russell H. (ed.): *Handbook and Tables in Science and Technology*, 2d ed., Oryx Press, Phoenix, 1983.
35. Bolz, Ray E., and George L. Tuve (eds.): *CRC Handbook of Tables for Applied Engineering Science*, CRC Press, Boca Raton, Fla., 1973.
36. Cheremisinoff, Nicholas P., and Paul N. Cheremisinoff: *Unit Conversions and Formulas Manual*, Ann Arbor Science Publishers, Ann Arbor, Mich., 1980.
37. Baumeister, Theodore (ed.): *Marks' Standard Handbook for Mechanical Engineers*, 8th ed., McGraw-Hill, New York, 1978.
38. Parmley, Robert O. (ed.): *Mechanical Components Handbook*, McGraw-Hill, New York, 1958. This handbook contains more technical data and computation than Part 2 of this book but covers fewer components.
39. Brady, George S., and Henry R. Clauser: *Materials Handbook*, 12th ed., McGraw-Hill, New York, 1985. This handbook describes many kinds of material in addition to the usual engineering metals and plastics.
40. Woodson, Wesley E.: *Human Factors Design Handbook*, McGraw-Hill, New York, 1981. This 1000-page book contains both encyclopedic text and design data.
41. McPartland, Joseph F. (ed.): *National Electrical Code Handbook*, 20th ed., McGraw-Hill, New York, 1990.
42. *Standard Industrial Classification Manual*, Executive Office of the President, Office of Management and Budget. For sale by National Technical Information Service, 5285 Port Royal Road, Springfield, Va., 22161; order No. PB87-100012. This book classifies almost *all* products into classes and subclasses and assigns a number to each class and subclass. The SIC number is used in many references to a company and its products and in the federal government's statistical reports.

20.5 Technology Encyclopedias

43. Meyers, Robert A. (ed.): *Encyclopedia of Physical Science and Technology*, 15 vols., Academic, New York, 1987.
44. *McGraw-Hill Encyclopedia of Science and Technology*, 6th ed., McGraw-Hill, New York, 1987.

20.6 Specifications and Standards

45. *NMTBA Specifications*, National Machine Tool Builders Association, 7901 Westpark Drive, McLean, Va. 22102. 703-893-2900.
46. *JIC Specifications*, National Machine Tool Builders Association, 7901 Westpark Drive, McLean, Va. 22102. 703-893-2900.

240   Topics in Design Engineering

47. *AGMA Standards,* American Gear Manufacturers Association, 1500 King Street, Suite 201, Alexandria, Va. 22314. 703-684-0211.
48. ASM specifications and other publications about metals, ASM International, Metals Park, Ohio 44073. 216-338-5151.
49. *NEMA Specifications,* National Electrical Manufacturers Association, 2101 L Street NW, Washington, D.C. 20037. 202-457-8400.
50. *Annual Book of ASTM Standards* (55,000 pages), American Society for Testing and Materials, 1916 Race Street, Philadelphia, Pa. 19103.
51. *Catalog of American National Standards,* Sales Department, American National Standards Institute, 1430 Broadway, New York, N.Y., 10018. 212-354-3300.
52. Department of Defense: *Index of Specifications and Standards (DOD-ISS),* Subscription Service, annually; Superintendent of Documents, Government Printing Office, Washington, D.C. 20402-9325.
53. *Index of Federal Specifications, Standards, and Commercial Item Descriptions,* U.S. General Services Administration, Government Printing Office, Washington, annually; nonmilitary.
54. *Military (MIL) Specifications,* Navy Publications and Forms Center, 5801 Tabor Avenue, Philadelphia, Pa. 19120. To purchase, telephone 215-697-3321.

20.6.1   Commercial sources for military specifications

55. Global Engineering Documents, 2625 Hickory Street, P.O. Box 2504, Santa Ana, Calif. 92707.
55. Information Handling Services, 15 Inverness Way, East Englewood, Colo. 80150. 800-525-7052.
56. National Standards Association, 5161 River Road, Bethesda, Md. 20816. 800-638-8076.
57. Information Marketing International, 13271 Northend Street, Oak Park, Mich. 48237. 313-546-6706.

20.7   Journals with New-Product Announcements

58. *Design News,* Cahners Publishing Co., 275 Washington St., Newton, Mass. 02158. 617-964-3030. (Free if you qualify.)
59. *Machine Design,* Penton Publishing Co., 1100 Superior Avenue, Cleveland, Ohio 44114. 216-696-7000. (Free if you qualify.)
60. *New Equipment Digest,* Penton Publishing Co., 1100 Superior Avenue, Cleveland, Ohio 44114. 216-696-7000. (Free if you qualify.)
61. *Materials Engineering,* Reinhold Publishing Co., 600 Summer Street, Stamford, Conn. 06904. 203-348-7531. Includes annual *Materials Selector Handbook.* (Free if you qualify.)

20.8   Directories

20.8.1   Purchasing directories

62. *Guide to American Directories,* B. Klein Publications, Coral Springs, Fla.
63. *Sweet's Mechanical Engineering Catalog,* Sweet's Catalogs, McGraw-Hill Book Company, 1221 Avenue of the Americas, New York, N.Y. 10020. 212-512-4442.
64. *MacRae's Blue Book,* MacRae's, 817 Broadway, New York, N.Y. 10003. 212-673-4700.
65. *Thomas Register of American Manufacturers,* Thomas Publishing Co., 1 Penn Plaza, New York, N.Y. 10110. 212-695-0500.

66. *Electronic Engineers' Master Catalog*, Hearst Business Communications Inc., 645 Stewart Avenue, Garden City, N.Y. 11530. 516-227-1300.

20.8.2 Business information about vendors

The following directories are useful primarily for business information about companies.

67. *Corporate Technology Directory*, Corporate Technology Information Services Inc., P.O. Box 81281, Wellesley Hills, Mass. 02181. 617-237-2001.
68. *U.S. Industrial Directory*, P.O. Box 10277, Stamford, Conn. 06904.
69. *Moody's Industrial Manual*, Moody's Investor Service Co., New York.
70. *Standard and Poor's Register of Corporations, Directors, and Executives*, Standard and Poor's Corp., New York.
71. *Dun & Bradstreet Directory and Reports*, Dun & Bradstreet, New York.

Your company controller or your bank can get you a detailed current report on any company, with information not published elsewhere. The reports emphasize credit status. References 69 and 70 are available in public libraries and in brokers' offices.

20.9 Mail-Order "Department Store" Catalogs

72. McMaster-Carr, P.O. Box 54960, Los Angeles, Calif. 90054. 213-945-2811.
73. Edmund Scientific Co., 4104 Edscorp Building, Barrington, N.J. 08007. 609-573-6266.
74. Newark Electronics, 4801 North Ravenswood Avenue, Chicago, Ill. 60640-4496. 312-784-5100.
75. Small Parts Co., P.O. Box 381966, Miami, Fla. 33238. 305-751-0856.
76. PIC Design Co., P.O. Box 1004, Benson Road, Middlebury, Conn. 06762. 203-758-8272.
77. Winfred M. Berg Inc., 499 Ocean Avenue, East Rockaway, N.Y. 11518. 516-599-5010.
78. Stock Drive Products, 2101 Jericho Turnpike, New Hyde Park, N.Y. 11040. 516-328-3300.
79. Value Plastics Inc., 3350 East Brook Drive, Fort Collins, Colo. 80525. 303-223-8306.
80. Boston Gear, 14 Hayward Street, Quincy, Mass. 02171. 800-343-3353.
81. L. A. Rubber Co., P.O. Box 23910, Los Angeles, Calif. 90023. 213-263-4131.
82. Linear Industries Inc., 1850 Enterprise Way, Monrovia, Calif. 91016. 818-303-1130.
83. W. W. Grainger Inc., 2738 Fulton Street, Chicago, Ill. 60612. 312-638-0536.
84. Minarik Electric Co., Suite 101, 165 East Commerce Drive, Schaumburg, Ill. 60173. 312-885-9337.

Index

Accuracy, 28
  (*See also* Repeatability)
Acme screws (*see* Screws)
Actuators, 135, 139
Adhesives, 66, 76, 154
Adjustable constraints, 63–69
  measurements, 66–69
  parameters, 63–65
  techniques, 65–66
Adjustment, 5
  screw, 15
Advertisements, 110
Air, compressed, 124
  (*See also* Pneumatics)
Air spring, 52
Airpot, 49
Anecdotes:
  accurate plumb line, 67
  angle generator, 58
  automatic screw thread gage, 87–91
  computer scientist vs. robot mechanisms, 234
  contract manufacturing, 163
  electrical connector, 43
  engineers' errors, 169
  explosive bolt cutter, 141
  flexure keyboard, 60–61
  heart-lung machine power, 214
  moon vehicle power, 213
  newspaper opener, 197–198
  robots for tank turrets, 177
  robots in China, 175
  rotation as lubricant, 153
  special hand tools, 173–174
  stress meter, 155
  vibration as lubricant, 153
  viscous clutch, 122
  wheel, invention of, 131
  windmill power, 211
Assembly, 3, 28, 174
  automatic, 172, 173
  deformation, 6, 7
  selective, 5
Automation:
  benefits of, 174
  classification of, 171–176
  optimum level of, 171–176
  policy questions, 175
Automobile:
  cylinder head, 17

Automobile (*Cont.*):
  differential, 12
  industry, 6
  transmission, 12
Axes, 8, 10
  robot, 180–181

Backlash, 27, 133, 215–221
  in bearings, 216–217
  in gears, 217
  in rack and pinion, 217–221
Ball and socket, 36
Bearings, 129–131
  air, 132–133
  ball, linear, 13, 24, 25
  ball, rotary, 13
  ball, self-aligning, 25, 79
  caster, ball, 87
  commercial, 116–119
  conical, 104
  flexure, 49
  hydrostatic, 45
  lubrication, 117
  roller, chained, 86
  roller, self-aligning, 86
  rolling, 116–118
  sliding, 118
    hydrostatic, 132
  spherical, 83
  spline, 87
  thrust, 45
Bellows, 56, 154
Belts, 133
Bibliography, 237–241
Big vs. small companies, 112
Bimetal, 59
Binding, 5, 6, 7, 14–15, 16, 27
  (*See also* Zero)
Blowers, 159
Bolt circle, 7, 17
Brain, 41
  (*See also* Humans)
Brakes (*see* Clutches and brakes)
Brand, 113
"Break in", 7
Bungee, 39

Caging, 32
Calibration, 7
Cam follower, 132

244 Index

Car (*see* Automobile)
Cartesian (*see* Axes; Robots)
Casters, 31, 132
Catalog:
 library, 110
 "mail-order", 241
Centering, 38–40
 hard, 38–39
 soft, 39–40, 42–45, 132
Centrifugal force, 10
Chain, 133
Chairs, 3, 17, 19–20
 (*See also* Stool)
Challenger, 98
Clad metal, 146
 (*See also* Bimetal)
Clamping, 17
Classification of components, 112, 113
Clutches and brakes, 37, 120–124
 control effects, 122–124
  angular position, 124
  centrifugal, 123
  electrical, 123
  human, 124
  torque, 123–124
 heat, 120
 jaw, 122
 lubricated, 121
 one-way, 121
 retarder, 120
 slip, 120
 torque effects:
  dry friction, 121
  eddy-current, 120, 122
  electric generators and motors, 122
  hydrodynamic, 121
  hysteresis, 120, 122
  magnetic particle, 121–122
  viscous drag, 120, 121
Collets, 120
Competition, electrical vs. mechanical engineering, 229–235
 electrical takeovers, 230–232
 future of mechanism, 232–235
 history, 229
Components, commercial, 107–159
 advantages, 107
 approved or preferred, 108–110
  company, 108–110
  military, 109
 benefits and limitations, 108–109
 blowers, 159
 classification, 112, 113

Components, commercial (*Cont.*):
 controls, 156–158
  computers, 157
  drum controllers, 156
  nonelectrical, 157–158
  programmable controllers (PLC), 157
  relays, 156–157
  timers, 156
 displays, 154–155
 enclosures, 146–147
 fasteners, 148–151
 guards, 154
 information sources, 111
 lamps, 158
 linear motion, 129–133
 lubrication, 153–154
 machine modules, 147
 magnets, 158
 nameplates, 158
 power sources, 135–143
 preferred product lists, 109
 pumps, 159
 rotary motion, 115–128
 seals, 154
 semifinished materials, 145–146
 sensors, 154–156
 shopping specification, 113
 springs, 10, 45, 152–153
 structural systems, 146
 tooling, 158
  (*See also* Jigs and fixtures)
 vibration and shock absorbers, 151–152
 washers, 159
Components, special, advantages, 107–108
Compressed air, 151
 (*See also* Pneumatics)
Constraints:
 area, 3
 ball and socket, 36
 ball and surface, 35
 bolted foot, 36
 centering, 38–40
  hard, 38–39
  soft, 39–40, 42–45
 cushioning, 43
 definition, 8
 fixed, 4
 forming, 30
 hard, 4, 35
 human, 40–42
  (*See also* Humans)

Index

Constraints (Cont.):
  point, 3
  point and surface, 35
  relative, 13
  roller and surface, 36
  rotary, 12, 31, 37
  shaft-and-sleeve bearing, 36
  soft, 4, 42–45, 46–55
    buoyancy, 47–48
    eddy currents, 48
    elasticity, 46
    fluid pressure, 49
    gravity, 48
    hysteresis, 46
    magnetism, 48
    viscosity, 46–47
  varying, 4
  vector, 11
  wheel, 36
    one flange on each of two wheels, 37
    pair of wheels on loose axle, 37
    pair of wheels on tight axle, 36
    railroad wheels, 37
    single narrow wheel, 36
    track following, 37
    two flanges on one wheel, 37
    V grooves, 37
  (See also Degree of freedom)
Consumers Union, 108
Contact:
  area, 7, 14
  line, 7
  point, 7
Controls, sequence, 156–158
  camshafts, 156, 158
  computers, 157
  drum controllers, 156
  fluidics, 157
  hydraulics, 157
  pneumatics, 157
  programmable controllers (PLC), 157
  punched tape, 158
  relays, 156
  timers, 156
Coordinates, 8
Cost, manufacturing, 3, 16
Coupling, flexible, 14
Coupling, phasing, 26
Cycloidal drives, 126
Cylinder, air, 31
Cylinder, hydraulic, 32
Cylinder, large thin wall, 7

Damage, 6
  (See also Deterioration)
Damping, 42, 52
Dashpots (see Vibration and shock absorbers)
Deflections, 30, 32
Deformation:
  assembly, 7
  elastic, 7
  load-spreading, 7
  operating, 7
  thermal, 6
  wear, 6
Degree of freedom, 8
Delrin, 117
Dent prevention, 43
Design books, 237–238
Deterioration, 225–228
Directories, 111, 112, 240–241
Displacement, 11
Displays, 154
Dissipation, energy, 51
Dovetail slide, 14–15, 131
Doweling, 6
Drawn shapes, 146
Drives:
  indexing, 127
  variable speed, 127–128
Dynamic braking, 127

Eccentrics, 66
Eddy currents, 51
  (See also Clutches and brakes)
Education, 170
Efficiency, 141
Elasticity, 133, 140
  (See also Flexures)
Elastomer pressure distribution, 43
Electro-magnetism, 10, 195–196
Electro-rheology, 50, 51, 121–122
Enclosures, 146
Encyclopedias, technology, 239
Energy storage, 62
  (See also Power sources)
Environment, 136, 198–199
ETL Testing Laboratories Inc., 109

Fasteners, 10, 41, 66, 159
  adjustment, 65
  clamping, 74–75
  latches, 62
  nonthreaded, 150–151
  retaining rings, 98–100, 119

Fasteners (Cont.):
  retention, 75–76
  as a seating force, 45
  threaded, 148–150
  V-band, 97–98, 119
  washers, unusual, 159
Feedback control, 136, 139, 155
Feet, 15, 36
Fixtures, 6, 21
  (See also Jigs and fixtures)
Flanged joint, 7, 17, 101
Flexible bodies, 32
Flexible constraints (see Flexures)
Flexible couplings, 57
Flexures, 55–63
  balancing spring, 62
  bimetal, 59
  clock, 62
  container, 62
  electrical, 59, 61
  energy storage, 62
  festoon, 60
  in Harmonic Drive, 125
  hose, 61
  keyboard, 60–61
  latch, 62
  musical instrument, 62
  pivots, 7, 49, 119
  polypropylene, 56
  ribbon cable, 60
  suspension, 58
  tape, 59
  tension, 61
  torsion, 58
Fluid pressure, 10, 45
Fluorolube, 47
Force, vector, 11
Forming constraints, 31
Freedoms, 8
Friction, 5, 11, 45, 51, 191–192
Friction constraints, 74–76
  belt and pulley, 75
  chucks and vises, 75
  clamps, 74
  collets, 74
  screw thread retention, 75
  setscrews, 75
  taper pins, 75
  variable speed drives, 75
  wedges, 75

Gasket, 17
Gears, 13, 125, 217

Geneva motion, 127
Gibs, 15
Gravity, 10, 45
Grippers, 191–204
  alignment and self-alignment, 200–204
  active, 203–204
  actuation, 199–200
  environment, 198–199
  gripping means, 191–198
  passive, 201–203
  (See also Robots)
Grout, 30–31, 66
Guards, 154
Gyroscope, 47, 58, 115

Handbooks, 238–239
Hard constraint, 51
Harmonic Drive, 125
Hertzian stress, 8, 35
Honeycomb, 146
Hose kink, 61
Hubs, coupling, 118
Humans, 40–42, 122, 210
  blood, 159
  control by, 172–174, 180
  and robots, 181, 188–189
  touch, hearing, and vision, 68–69
Hydraulics, 123, 138
Hype, 223–224
Hysteresis, 51

Indexing drives, 127
Industrial engineering, 167
Inspection machines:
  large, 31
  screw thread, 87–91
Instrumentation, 25

Jamming, 24
Japanese, 170
Jigs and fixtures, 21, 22, 145, 158
Journalism, 224
Justification, cost, 169, 174

Kingsbury thrust bearings, 45, 117

Labor, 3
Laminates, 146
Lamps, 158
Lapping, 101–104
Large loads, 129
Latches, 151
Lathe chuck, 23

Index

Lead screws, 16, 116, 133
   acme, 131, 133
   ball-bearing, 133, 196
   (*See also* Spline)
Linear-motion mechanisms, 24–25
Linear slide, 130
Load dividers, 32–34
   architectural, 33
   gravity, 34
   linkage, 32–34
   sedan chair, 34
   telescope, 34
   whiffletree, 32, 34
Load spreading, 7
Loads, 6
   heavy, 7
Looseness, 3, 5, 6, 7, 14–15, 16, 27
   (*See also* Zero)
Lubrication, 12, 44–45, 117, 121, 140, 153–154
   dry lubrication, 153

Machine modules:
   robots, 147
   slides, 147
Machine tools, 21, 32
Machining, 27, 28
Magnetism, 10, 45
Magnets, 158
Maintenance, 168
"Make or buy", 163
Management, manufacturing engineering, 168–169
Manufacturability, 171
Manufacturing:
   lead screw, 16
   matched, 5
   processes, 163–166
   (*See also* Doweling; Jigs and fixtures; Tolerances)
Manufacturing engineering, 167–170
   Hewlett-Packard, 169
   IBM, 169
   3M, 169
Matched sets, 12–13, 93–94
   as RedCD, 93
   rolling bearings, 129–130
   (*See also* Bearings, rolling)
   wheel pairs, 26, 27
   X-Y table, 130–132
Materials, soft constraint, 46
Mechanical filter, 52
Mechanical phase angle, 63, 66

Mechanism designers, 171
Mechanization:
   classification, 171–176
   optimum level, 171–176
MinCD (*see* Minimum constraint design)
Minimum constraint design (MinCD), 52, 133
   cart, 31
   comparison with semi-MinCD and RedCD, 17
   definition, 3, 9
   examples, 9, 17–34
   flexible bodies, 30, 32
   jigs and fixtures, 21
   lead screw, 16, 70
   long machine beds, 30
   purity, 9, 10–11, 13, 26–27, 45, 49, 93
   theory of, 8–9, 14, 16
   tripods, 17–23, 41, 129
Modularity, 183
Modules, machine:
   dial tables, 147
   linear transfer tables, 147
   part feeders, 147
   pick-and-place mechanisms, 147
   (*See also* Pick and place)
   slides, 147
   spindles, 147
   ways, machine, 14–15
   (*See also* Robots)
Monorails, 130
Motivation, 169
Motors, electric:
   dc, 72
   induction, 72
   linear, 72
   servo, 72
   stepping, 72
   synchronous, 72
   (*See also* Feedback control; Power sources)

Nameplates, 158–159
NBS Handbook H28, 89, 148
NEMA, 146
Nylon, 117

Orientation, 11
   (*See also* Axes)
Overconstraint, 11, 14, 15, 23, 34, 35
   (*See also* Redundant constraint design)
Overloads, 6
Overtravel, 43

Overtravel cushioning, 43
Overturning, 3
  (*See also* Stability)

Perforated metal, 145
Piano hinge, 101
Pick and place, 127
  (*See also* Assembly; Robots)
Pinion rod, 146
Pitch (*see* Axes)
Plumbing, 139–140
Pneumatics, 151
  (*See also* Power sources)
Polaroid, 68
Polyurethane tires, 26, 43, 131
Powder-actuated devices (*see* Power sources)
Power sources, 62, 135–143
  animals, 143, 210–211
  elasticity, 212–213
  electricity, 135–138, 206–207
  explosives, 141, 210
  flywheels, 142
  fuel burning, 142, 211
  gravity, 212
  heat engines, 142, 209–210
  humans, 135, 210
    (*See also* Humans)
  hydraulics, 138–140, 208–209
  inertia, 213
  nuclear, 214
  pneumatics, 140–141
  springs, 141
    (*See also* Springs)
  sunlight, 211
  utility water, 213–214
  vacuum, 209
  wind, 211–212
Pumps, 159

R&D, 168
Rack and pinion, 217–221
  (*See also* Gears)
Rectangular ways, 14–15
RedCD (*see* Redundant constraint design)
Redundant constraint design (RedCD), 93
  bad, 14–17
  components, 97–101
    retaining rings, 98–99
    screw threads, 100
    V-band fastener, 97–98, 119
  definition, 3
  disadvantages, 5–7

Redundant constraint design (RedCD) (*Cont.*):
  example, 28
  good, 17
  human, 41
  lead screws, 70
  matched set, 12
    (*See also* Matched sets)
  self-improving, 101–104
  useful, 93–104
    large distributed loads, 95
    necessary deformation, 96
    varying load distribution, 96–97
References, 237–241
Relays, 146
Reliability, 3
Repeatability, 28
  (*See also* Accuracy)
Restoring force (*see* Seating force)
Retaining rings, 98–100, 119
Retarders, 51
Robots, 28, 147, 172, 177–190
  accuracy, 186–187
  Cartesian, 182–185
  comparison with humans, 181, 188–189
  configurations, 184–185
  controls, 179–180, 182
    computer, 182
    PLC, 182
    (*See also* Controls, sequence)
  conversion, task to task, 187
  cost justification, 187–188, 189–190
  end effectors, 178, 179, 191–204
  grippers, 23–24, 53, 133
    (*See also* Grippers)
  history, 177–178
  mechanisms, 180–185
  myth and reality, 177–178
  programming, 185–186
  repeatability, 186–187
  safety, 184
  spar, 31
  storage and retrieval, 26–27
  uses, 178–179
Roll (*see* Axes)
Rollers, 13, 129–132
Rope, 133
Rotary union, 123, 154
Rotation, 14
Rotation components:
  belts, 125–126
  chain, 126

Rotation components (*Cont.*):
  flexible couplings, 124
  gears, 125
  universal joints, 124

Safety caging, 28
Salespeople, 111
Scratch prevention, 43
Screws:
  lead screws, 16, 116, 120, 133
    acme, 131, 133
    ball, 133, 196
  as RedCD, 100
  retention, 75–76
    adhesives, 76
    deformed threads, 75
    locknuts, 76
    lockwashers, 76
    plastic inserts, 75
    self-tapping, 75
    tapered threads, 75
  specifications, 89
  (*See also* Fasteners)
Seals, 154
Seating force:
  antibacklash, 215
  cylinders, 51
  examples, 10, 24, 25–26
  explanation, 8, 10, 42
  gravity, 48
  sources, 45–46
Self-alignment, 25, 76–91
  active, 91
  ball bearings, 79, 83
  caster, 18, 76–77
  gimbals, 77–79
  spherical (ball) joint, 79–81
    leveling pad, 82
    spherical washers, 85–86
  universal joint, 79
  (*See also* Robots, grippers)
Semi-MinCD (*see* Semi-minimum constraint design)
Semi-minimum constraint design (semi-MinCD), 93
  bolted foot, 95
  conversion from MinCD, 94–95
  definition, 4
  heavy loads, 94
  lathe carriage, 94
  matched sets, 13
    (*See also* Matched sets)
  zero-looseness hinge, 95

Sensors, 41, 154
Servo (*see* Feedback control)
Setscrews, 118
Shafting, 119–120, 145
  flexible, 120
  splined, 120
  threaded, 120
Shapes, rolled and extruded, 145
Shims, 66
Shock, 42
Shock mounts, 52
Silicone, 47, 49
Silly Putty, 47
Skill, 28
Slip, 51
Society of Manufacturing Engineers, 169
Soft constraint, 51
  brakes, 51
  clutches, 51
  cylinders, 50–51
  dashpots, 49–50
  inflated devices:
    actuators, 55
    air springs, 52
    expanding O rings, 53
    grippers, 53
    seal, 52
    squeeze valve, 54
    tube, 53, 54
    vacuum grippers, 53
    ventricles, 54
  motors, 51
    (*See also* Motors, electric)
  shock mounts, 51–52
  voice coils, 51
Space shuttle, 96
Special machines, 168
Specifications and standards, 89, 148, 239–240
Speed reducers, gearless, 125
Spindle assemblies, 118, 147
Spline, 14, 37, 38
Springs, 10, 45, 152–153
Stability, 11–12, 37, 41
Static determinacy, 11
Steam engines (*see* Power sources)
Steering:
  belt, 73–74
  tricycle, 22
  wheel, 132
Stiction, 49
Stool, 3, 18–19
  (*See also* Chairs)

Stresses, 6, 32
  contact, 43
  Hertzian, 8
  meter, 155
Structural systems, 146
Surface plate, 21

Tables, 17
Tandem shafts, 29–30
Tapers, 119
Technician work, 168, 169
Teflon, 117
Thixotropy, 47
Thomas Register, 147, 240
Thread gage, automatic, 12, 87–91
Tightness, 5
  (*See also* Zero)
Tolerances:
  in load divider, 32
  manufacturing, 3, 5, 28
Tooling components, 158
  (*See also* Jigs and fixtures)
Tooling plate, 145
Tracks, 129–132
Trade shows, 110–111
Trademarks, 113
Trailer, 23
  hitch, 28
Transducers, 69, 154
Tricycle, 22
Tripods, 17–23, 41
Tubing, 145

Underwriters Laboratories Inc., 109
Universal joints, 14, 38, 124
  flexure, 124

V bands, 97–98, 119
Vacuum, 192–194
Valves, 123, 139
Variable constraints, 69–74
  belt and pulley, 73–74
  cams, 69
  chain, tape, and rope, 73
  cylinders, fluid pressure, 70
  gears, 72
  lead screws, 70
  linkages, 70
  motors, electric, 71–72
  rack and pinion, 72

Variable constraints (*Cont.*):
  servos, 74
  (*See also* Feedback control; Hydraulics)
Variable-speed drives, 127–128
  electric motor, 127
  engines, 128
  friction, 128
  gearshift, 128
  hydraulic, 128
  slip clutch, 127
  V belt, 128
Vehicles:
  bicycle, 41
  pogo stick, 42
  skateboard, 41
  train, 48
  tricycle, 22
  unicycle, 42
  wheelbarrow, 41
  (*See also* Automobile)
Vendors, specialized, 163
Vertistat, 49
Vibration, 42
Vibration and shock absorbers, 151–152
Viscosity, 45, 49, 51, 121

Washers, unusual, 159
Ways:
  dovetail, 131
  rectangular, 131
"Wear in", 7, 12, 101–104
Wedges, 66
Wheels, 13, 129, 131–132
Whiffletrees, 32, 34
  (*See also* Load dividers)
Wood paneling, 146

X, Y, Z (*see* Axes)
X-Y table, 131

Yaw (*see* Axes)
Zero:
  in backlash, 57
  in binding, 3
  in friction, 58
  in looseness, 3, 55
    hinge, 95
  in lubrication, 55
  in stiction, 55
  (*See also* Flexures)